STUDY GUIDE

PETER O. MULLER
ELIZABETH MULLER HAMES
University of Miami

Geography: Realms
Regions
and Concepts
2000

NINTH EDITION

H. J. de Blij
Marshall University

Peter O. Muller
University of Miami

JOHN WILEY & SONS, INC.
New York • Chichester • Weinheim • Brisbane • Singapore • Toronto

To order books or for customer service call 1-800-CALL-WILEY (225-5945).

ISBN 0-471-35748-0

Printed in the United States of America

10 9 8 7 6 5 4 3 2

Printed and bound by Courier Kendallville, Inc.

A FOREWORD TO STUDENTS

This **Study Guide** is designed to enhance your learning of world regional geography, building upon your parallel experiences in the classroom and in reading the chapters of the textbook. It is recommended that you have an *atlas* handy; the base maps in the textbook will readily suffice if no other atlas is available. The only other resources you will need to complete the exercises in this manual are a set of colored pencils or pens and (optional) tracing paper.

One of the main purposes of this **Study Guide** is to get you to become more familiar with maps (a subject introduced in the textbook's Appendix A). For each world geographic realm, we have provided four outline maps (located at the end of each chapter) on which you are asked to locate and enter important geographic information. The essay questions in each chapter are also attuned to maps—many, in fact, cry out for a cartographic answer. That is exactly what you should consider doing: use sketch maps whenever the map is the best way to explain what you mean (like a picture, a map is often worth a thousand words). You can use tracings of the printed maps in order to get the necessary outlines of the geographic realms; it is also a good idea to use such tracings in answering map questions as a sort of "test run" before placing your final answers on the printed map. Moreover, placing information on maps as you read the textbook can be a valuable technique for later reviewing before examinations; an extra printed outline map is always provided in each chapter of this manual. This map of Southern Africa shows what a typical review map might look like after basic geographic information from your reading and **Study Guide** exercises is added:

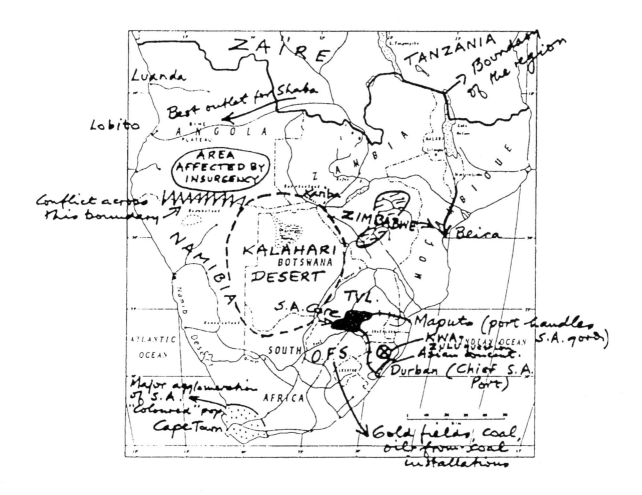

Each chapter of this **Study Guide** follows an identical format. The opening section describes the *objectives* of the chapter together with a list of things you should be able to do once you have learned the content of the parallel textbook chapter. This is followed by a *glossary* of terms covered in the book chapter, including appropriate terminology from the main textbook glossary as well as additional definitions; the **Study Guide** glossary is arranged in the order in which terms appear in the book, and carries page references to the text. Then comes the main study and review section, a series of detailed, side-by-side *questions and answers* for you to sharpen your understanding of the high points of the subject matter. *Map exercises* appear next, divided into two sections: (1) a series of map-comparison queries arranged around the textbook maps, and (2) three assignments to fill in basic physical, political-cultural, and economic-urban geographic information for each realm; as indicated above, a fourth blank map is included for your own use. After this comes a *practice examination* consisting of short answers (multiple choice, true-false, fill-ins, and matching) and some longer essay questions—an answer key for the short-answer questions is found at the back of this manual. Finally, a *term paper pointers* section follows that includes tips on the preparation of research papers on the realm covered in the book chapter, and is built around a guide to the literature cited in each chapter bibliography (located in the References and Further Readings section at the back of the textbook).

You should find these exercises helpful in this course, particularly for personal studying purposes. They will also assist you to learn basic world geography and to sharpen your map interpretation skills. Such knowledge is not only important as a contribution to your becoming a liberally-educated person—it may already be vital for success in the ever more globalized economy and society of which we are now all a part.

A FOREWORD TO INSTRUCTORS

The aims of this Student Supplement to the ninth edition of *Geography: Realms, Regions, and Concepts* are spelled out in the Student Foreword above.

This student **Study Guide** is designed to complement the text in several ways, with each chapter providing a lengthy glossary, a detailed list of study questions-and-answers, map exercises that require comparisons and placement of spatial information on outline maps, practice examination questions, and term paper pointers. You should also be aware of the availability of the **Instructor's Manual** (*as well as other ancillaries listed in the textbook's Preface*) to the ninth edition, which contains a number of items pertinent to the teaching of a world regional geography course based on this textbook.

ACKNOWLEDGMENTS

We are most grateful to a number of people for their assistance and advice during the preparation of this **Study Guide**. The manuscript was produced in camera-ready form by Karen Tina Sheskin and Prof. Ira M. Sheskin, Department of Geography, University of Miami. The maps were designed by Mapping Specialists, Inc. of Madison, Wisconsin. At the department office at the University of Miami, Hilde Al-Mashat and Dresdin Ferrer and their student assistants performed countless supporting tasks with their usual efficiency and cheerful spirits. At Wiley we are most thankful for the support and assistance of Alicia Solis, Nanette Kauffman, Edward Starr, Karen Ayoub, and Barbara Bredenko. We also acknowledge again those who helped in the preparation of the four most recent editions of this **Study Guide**—on which this latest version is based.

Peter O. Muller
Elizabeth Muller Hames
Coral Gables, Florida
May 24, 1999

TABLE OF CONTENTS

INTRODUCTION
WORLD REGIONAL GEOGRAPHY

OBJECTIVES OF THIS CHAPTER

The Introduction is an overview of world regional geography. Proceeding from a discussion of basic concepts of regionalism, the topics of scale, the natural environment, culture, landscape, and global population patterns are explored. Additional human-geographic variables are introduced, notably political and economic forces (the latter are explored in some depth within the context of economic development), and provide the basis for establishing the framework of world geographic realms to be pursued in this book. Following capsule previews of each realm, the relationship between regional and systematic (topical) geography is considered. A final box introduces you to the discipline and practice of geography, a prelude to an essay in Appendix B that examines career opportunities in this expanding, wide-ranging field.

Having learned this basic background to the subject, you should be able to:

1. Differentiate among the various basic concepts of regions, and understand the nature of regional structure.

2. Understand the general aspects of the natural environment and its evolution over the past few million years.

3. Explain broad, world-scale patterns of terrain, hydrography (water distribution), and climate, with some emphasis on the interrelationship of these environmental systems.

4. Understand the notion of culture and how it is expressed to form human landscapes in time and space.

5. Identify and discuss the global distribution of humankind, with emphasis on the leading population concentrations.

6. Understand the importance of influential developed areas and how, at the global scale, the economically advantaged countries function as a core that dominates the less developed countries that constitute a "have-not," or economically-disadvantaged periphery.

7. Name, map, and differentiate among the 12 world geographic realms that form the spatial framework to be followed in the remaining chapters of the book.

8. Read and interpret the maps that accompany the text, knowing such map properties as scale, projection, orientation, and point, line, and area symbols (Appendix A—Map Reading and Interpretation—covers most of these points, and should be considered a supplement to this Introductory chapter).

GLOSSARY

Geographic realm (2)

The basic spatial unit in our world regionalization scheme. Each realm is defined in terms of a synthesis of its total human geography—a composite of its leading cultural, economic, historical, political, and appropriate environmental features.

Spatial perspective (2)

Pertaining to space on the Earth's surface; synonym for *geographic(al)*.

Taxonomy (2)

A system of scientific classification.

Transition zone (3)

An area of spatial change where the peripheries of two adjacent realms or regions join; marked by a gradual shift (rather than a sharp break) in the characteristics that distinguish these neighboring geographic entities from one another.

Region (5)

An area of the Earth's surface marked by specific criteria. A commonly used term and a geographic concept of central importance.

Regional boundaries (6)

In theory, the line that circumscribes a region. But razor-sharp lines are seldom encountered, even in nature (e.g., a coastline constantly changes depending upon the tide).

Location (6)

The position of a place or region on the Earth's surface. *Absolute location* is that position expressed in degrees, minutes, and seconds of latitude and longitude; *relative location* is that position relative to the position of other places and regions. Thus Palmyra, New Jersey is located at 40° North latitude, 75° West longitude; relatively speaking, this suburban community is located on the eastern bank of the Delaware River facing Tacony, a factory neighborhood in the Lower Northeast portion of the city of Philadelphia, Pennsylvania.

Formal region (6-7)

A type of region marked by a certain degree of homogeneity in one or more phenomena, such as physical or cultural properties, or both.

Functional region (7)

A region marked less by its sameness than its dynamic internal structure, formed by a set of places and their interactions.

Spatial system (7)

Any group of objects or institutions and their mutual interactions; geography treats systems that are expressed spatially—such as regions. *Functional regions* are spatial systems.

Hinterland (7)

The surrounding area served by an urban center. This area is both served by the urban central core and comes under its cultural and economic influence.

Scale (7)

Representation of a real-world phenomenon at a certain level of reduction or generalization. In cartography, the ratio of map distance to actual ground distance; shown on map as bar graph, representative fraction, and/or verbal statement.

Natural landscape (8)

The array of landforms that constitutes the Earth's surface (mountains, hills, plains, and plateaus) and the physical features that mark them (such as water bodies, soils, and vegetation). Each geographic realm has its distinctive combination of natural landscapes.

Physical geography (9)

The spatial study of the Earth's natural phenomena and their systems, processes, and structures.

Continental drift (9)

The slow movement of continents controlled by the processes associated with *plate tectonics*.

Plate tectonics (9-10)

Bonded portions of the Earth's mantle and crust, called *tectonic plates*, averaging 60 miles (100 kilometers) in thickness. More than a dozen such plates exist (see Fig. I-4), most of continental proportions, and they are in motion. Where they meet one slides under the other, crumpling the surface crust and producing significant volcanic and seismic (earthquake) activity. A major mountain building force.

Subduction (10)

In *plate tectonics*, the process that occurs when an oceanic plate converges head-on with a plate carrying a continental landmass at its leading edge. The lighter continental plate overrides the denser oceanic plate and pushes it downward.

Pacific Ring of Fire (11)

Zone of crustal instability along tectonic plate boundaries, marked by earthquakes and volcanic activity, that rings the Pacific Ocean basin.

Desertification (11)

The process of desert expansion into neighboring steppelands as a result of human degradation of fragile semiarid environments.

Ice age (12)

A stretch of geologic time during which the Earth's average atmospheric temperature is lowered; causes the equatorward expansion of continental icesheets in the high latitudes and the growth of mountain glaciers in and around the highlands of the lower latitudes.

Pleistocene epoch (12)

Recent period of geologic time that spans the rise of humanity, beginning about 2 million years ago. Marked by *glaciations* (repeated advances of continental-scale icesheets) and milder *interglaciations* (ice withdrawals). Although the last 10,000 years are known as the Holocene, Pleistocene-like conditions seem to be continuing and the present is likely to be another Pleistocene interglaciation; the glaciers will return.

Hydrologic cycle (13)

The hydrosphere, or Earth's vital supply of water, ice, and water vapor, is maintained by the *hydrologic cycle*, which circulates water in a constant repetitious cycle of evaporation, condensation, precipitation, and runoff.

Climate (11-13)

The long-term conditions (over at least 30 years) of aggregate weather over a region, summarized by averages and measures of variability; a synthesis of the succession of weather events we have learned to expect at any given location.

Physiography (16)

The total physical geography that forms the natural-environmental base of a realm or region.

Culture *(16)*

According to Ralph Linton et al., "the sum total of knowledge, attitudes, and habitual behavior patterns shared and transmitted by members of a society." Of course, this is but a representative definition because no single one can do the concept full justice.

Cultural landscape *(17-18)*

The composite of human imprints on the surface of the Earth. By this progressive imprinting of the human presence, the physical landscape is modified into the cultural landscape, forming an interacting unity between the two. May also be regarded as a composition of artificial spaces that serves as background for the collective human experience.

Cultural process *(17)*

Causal force that shapes a cultural spatial pattern or structure as it unfolds over time.

Sequent occupance *(17-18)*

The idea that successive societies leave their cultural imprints on a place, each contributing to the cumulative cultural landscape.

Central Business District (CBD) *(18)*

The downtown heart of a central city, the CBD is marked by high land values, a concentration of business and commerce, and the clustering of the tallest buildings.

Ethnicity *(18)*

The combination of a people's culture (traditions, customs, language, and religion) and racial ancestry.

Population distribution *(18)*

The way people have arranged themselves in geographic space. One of human geography's most essential expressions because it represents the sum total of the adjustments that a population has made to its natural, cultural, and economic environments.

Population density *(19)*

The number of people per unit area.

Megalopolis (20)

When spelled with a lower-case m, a synonym for *conurbation*, one of the large coalescing supercities forming in diverse parts of the world. When capitalized, refers specifically to the multi-metropolitan corridor that extends along the northeastern U.S. seaboard from north of Boston to south of Washington, D.C.

Cartogram (20)

A specially transformed map not based on traditional representations of scale or area.

Urbanization (20-21)

A term with several connotations. The proportion of a country's population living in cities and towns (metropolitan areas) is its level of urbanization.

State (22)

A politically organized territory that is administered by a sovereign government and is recognized by a significant portion of the international community. A state must also contain a permanent resident population, an organized economy, and a functioning internal circulation system.

European state model (23)

A state consisting of legally defined territory, inhabited by a population governed from a capital city by a representative government.

Development (23-24)

The economic, social, and institutional growth of national states. This can be measured in a number of ways; the text uses the fourfold scheme for ranking countries according to national income devised by the World Bank (Fig. I-12; pp. 26-27). Other leading development indicators are: national product per person, labor-force occupational structure, productivity per worker, per capita energy consumption, transportation/communications facilities per person, manufactured-metals consumption per person, and such rates as literacy, per capita savings, and individual caloric intake.

Economic geography (23)

The field of geography that focuses on the diverse ways in which people earn a living, and how the goods and services they produce are expressed and organized spatially.

Core area *(28-box)*

In geography, a term with several connotations. Core refers to the center, heart, or focus. The core area of a state is constituted by the national heartland, the largest population cluster, the most productive region, and the part of the country with the greatest centrality and accessibility— probably containing the capital city as well.

Regional disparity *(25)*

The spatial unevenness in standard of living that occurs within a country, whose "average," overall income statistics invariably mask the differences that exist between the extremes of the wealthy core and the poor periphery.

Neocolonialism *(26-27)*

The term used by developing countries to underscore that the entrenched colonial system of international exchange and capital flow has not changed in the postcolonial era—thereby perpetuating the huge economic advantages of the developed world.

Geographic realm *(29-32)*

The basic spatial unit in our regionalization scheme. Each realm is defined in terms of a synthesis of its total human geography—a composite of its leading cultural, economic, historical, political, and appropriate environmental features.

Globalization *(32)*

The gradual reduction of regional contrasts resulting from increasing international cultural, economic, and political exchanges.

Regional iconography *(33)*

A region's unifying symbols and habits of culture and tradition.

Systematic geography *(34)*

Topical geography: cultural, political, economic geography, and the like.

SELF-TESTING QUESTIONS

Cover the right side of the page with a sheet of paper. Uncover each line after you have attempted to answer the question in the left column. If necessary, refer to textbook page(s) listed at the right.

Question	Answer	Page
Regional Concepts		
Define *spatial*.	Pertaining to space on the Earth's surface; synonym for "geographic(al)."	2
What does the term *hierarchy* mean?	A rank-ordering of phenomena, with each level subordinate to the one above it and superior to the one below.	2
What characterizes each major world realm?	A special combination of cultural, physical, historical, economic, and organizational qualities.	3
Explain how regionalization is the geographer's means of classification.	According to selected criteria, specific meanings are assigned to certain areas.	3
How do most regional boundaries appear in the landscape?	As transition zones rather than razor-sharp lines.	3
Why is *relative location* a more practical measure than *absolute location*?	*Absolute location* refers only to latitude and longitude, whereas *relative location* refers to a place in terms of its position relative to other regions.	6
How does a *formal region* differ from a *functional region*?	*Formal regions* are marked by internal homogeneity; *functional regions* are defined by their structuring as spatial systems.	6-7
How can certain regions be conceptualized as systems?	Non-uniform regions are marked by a set of integrated activities that interconnect their various parts.	7
Concepts of Scale		
What is *scale*?	Broadly, the level of generalization; specifically, the ratio of map distance to actual ground distance.	7

What is a *representative fraction*?	The fraction specifying the scale of map; an *RF* of 1:63,360 stands for 1 map inch representing 63,360 ground inches (or exactly 1 mile).	7

Environmental Change

Do all of the continents contain shields?	Yes, each of the continents have an old geologic core, or shield.	9
What is meant by the term *plate tectonics*?	The recently proven theory that the Earth's crust consists of a set of rigid plates that are in motion, producing great stresses at their boundaries.	9-10
What is the Pleistocene?	A recent geologic epoch (2 million to 10,000 years ago) marked by successive glaciations and interglaciations.	12
What are interglaciations?	The relative warm spells that separate glaciations.	12

Hydrography (Water)

What are the components of the *hydrologic cycle*?	Evaporating moisture from water bodies, condensation, precipitation, and surface runoff.	13
Why are equatorial and tropical areas generally well-watered locations?	The hydrologic cycle is most efficient in high heat and humidity—carrying large amounts of evaporated ocean water to fall as rain.	13
Is moisture evenly distributed across the globe?	Most decidedly not; the text discussion and Fig. I-7 cover the details.	12-16

Climatic Regions

What does the world regional pattern of climates look like?	The detailed distribution is mapped in Fig. I-8.	14-15
What are the general features of *A* climates?	These humid tropical climates are high-heat and high-moisture; important subtypes are rainforest, savanna, and monsoon climates.	14-15
What are the general features of *B* climates?	Dryness is dominant, and they occur in lower, middle, and higher latitudes; semiarid steppe and arid desert are the major subtypes.	15

What are the general features of *C* climates?	Humid, temperate, mid-latitude climates; found in eastern U.S., Western Europe, and elsewhere; Mediterranean climate is a major subtype.	15-16
What are the general features of *D* climates?	Humid cold climates most prominent in the higher-latitude interiors of large Northern Hemisphere landmasses; huge annual temperature ranges.	16
What are the general features of *E* climates?	Cold polar climates, differentiated into tundra and icecap subtypes.	16
What are the *H* climates?	Undifferentiated high-altitude climates of mountains at any latitude; much like the *E* climates.	16

Cultural Landscape

How can humans leave their imprint on the land surface?	In numerous ways, because people are agents of change; their structures and artifacts progressively transform natural into cultural landscapes.	17
What exactly is the *cultural landscape*?	According to Carl Sauer, the composite imprints and forms superimposed on the physical landscape by human activities.	17
What is meant by *sequent occupance*?	The successive stages in the evolution of a region's cultural landscape.	17-18

Space and Population

How large is the population of the world, and what are the immediate prospects for growth?	In 2000 the total was 6.1 billion; at present growth rates, a net gain of approximately 85 million people is added every year.	18
What does the global distribution of humanity look like?	See Fig. I-9 (pp. 20-21).	19-21
Where are the four largest population agglomerations located?	In descending size, East Asia, South Asia, Europe, and Eastern North America.	19-20
What are the main features of the East Asian population cluster?	China's Huang He and Chang Jiang river valleys; dominantly rural population.	19

What are the main features of the South Asian population cluster?	India's Ganges Lowland, Pakistan's Indus Valley, and all of Bangladesh in the Ganges-Brahmaputra Delta; heavily rural.	19
What are the main features of the European population cluster?	A central east-west population axis, oriented to industrial resources and highly urbanized.	19
What are the main features of the North American population cluster?	A pattern like the European cluster, dominated by (the Boston-Washington) Megalopolis along the northeastern U.S. seaboard.	19-20
Which of the major population agglomerations is likely to be the largest a century from now?	At current growth rates, South Asia will overtake East Asia as the world's greatest population cluster during the 21st century.	19

Political Geography

How many political units are there today?	In the year 2000, more than 180 national states.	23
What is the *European state model*?	A state with a legally defined territory, inhabited by a population governed from a capital city by a representative government.	23

Economic Geography

What is the central concern of economic geography?	Studying the varied ways in which people earn a living, and how the goods and services they produce are spatially distributed and organized.	23
What are the four economic groups that countries are divided into?	High income, upper-middle income, lower-middle income, and low income. The global pattern of these countries is mapped in Fig.I-12.	23-24

Patterns of Development

Why is the world no longer simply divided into "developed" and "undeveloped" countries?	Many countries, despite their overall level of income, display cores of development that resemble rich societies, and peripheries of extreme poverty.	24-25

| How is economic development measured? | According to a number of indices that facilitate international comparison; Table I-1 lists some of the most commonly used variables. | 25-26 |
| What are some of the obstacles that under-privileged areas face? | Beyond a disadvantageous position in the global economic system, less advantaged countries exhibit political instability, corrupt leadership, misdirected priorities, misused aid, and traditionalism. | 27-28 |

Geographic Realms

What is a *geographic realm*?	A geographic realm is a large regional unit based on broad multiple criteria—reflecting physical, economic, political, urban, historical, and population as well as cultural geography.	29-32
What does the distribution of world geographic realms look like?	This framework, which underlies the organization of the rest of this book, are mapped in two levels of detail in Figs. I-2 and I-13; both maps should be studied carefully.	4-5; 30-31
What are the names of the 12 geographic realms?	Europe, Russia, North America, Middle America, South America, North Africa/Southwest Asia, Subsaharan Africa, South Asia, East Asia, Southeast Asia, the Austral Realm, and the Pacific Realm.	29-32
What are the major characteristics of each realm?	They are too numerous to list here; study pp. 29-32 closely—the matching question in this chapter of the **Study Guide** will also be helpful to you.	29-32

Regional and Systematic Geography

| What are the *systematic* subfields of geography? | The topical branches of the discipline that cut across the world realms, such as political or economic geography; the systematic subfields are diagramed in Fig. I-14 on p. 34. | 34 |

MAP EXERCISES

Map Comparison

1.   Comparing Figs. I-5 (p. 10) and I-4 (p. 9), what generalizations can be made about the landforms that are located along tectonic plate boundaries, especially where plates are pushing against one another?

2.   Comparing Figs. I-5 (p. 10) and I-6 (p. 11), what generalizations can be made about the location of Pleistocene icesheets in the Northern Hemisphere and the distribution of mountain ranges?

3.   On Fig. I-9 (pp. 20-21), compare and contrast the *internal* patterns of the world's largest population agglomerations. Can some of the variations be accounted for by differences in environmental patterns as mapped in Figs. I-7, and I-8?

4.   Compare the advantages and disadvantages of generalizing world climate-region patterns (Fig. I-8 on pp. 14-15), with special emphasis on the mainland United States.

5.   Carefully compare the map of world political units with the world population distribution map (Figs. I-11 and I-9, respectively). List five countries that are prominent in size on the population map but not on the world political map, and five countries that are large in territorial size on the political map but modest in appearance in terms of population.

6.   Read Appendix A (Map Reading and Map Interpretation). Utilizing Appendix A as a guide, review all of the maps in the Introduction chapter in order to find examples of point and line symbols (area symbols, of course, abound in these maps).

7.   Compare and contrast the Cartogram of the World's National Populations (Fig. I-10, p.22), and the map of the States of the World, 2000 (Fig. I-11, pp.24-25). In what ways is each view a distortion? Does the world view appear markedly different when countries are represented in population space? Which nations appear to dominate in size on one map, but not the other?

Map Construction   *(Use outline maps at the end of this chapter)*

1.   In order to examine and study the interrelationships among environmental systems, draw in the following for each of the world's continents:

   a.   in *light red*, sketch in all major highland areas (use Fig. I-1 as a source, which is found on the pair of pages preceding p. 2 in the textbook).

b.   in *green*, sketch in the wettest zones (the color closest to the top of the legend color-box in Fig. I-7 on p.12); in *yellow*, sketch in the driest zones (color closest to the bottom of the color-box).

c.   in *black* or *blue*, draw in all climate-zone boundaries from Fig. I-8 (pp. 14-15) and label each climate region with its appropriate Köppen-Geiger letter symbol.

2.   Using Fig. I-9 (pp. 20-21) as a guide, draw in and label each of the major world population agglomerations discussed on pp. 18-20 in the text.

3.   Using Fig. I-13 (pp. 30-31) as a guide, draw in the boundaries of all of the 12 world geographic realms and label each realm; *you should be able to do this by memory if you have learned Fig. I-13 first.*

PRACTICE EXAMINATION

Short-Answer Questions

Multiple Choice

1.   Which of the following cartographic devices when deployed on a map will tell you its scale:

a) longitude   b) legend   c) representative fraction
d) color key   e) latitude

2.   The islands of the Caribbean Sea belong to which of the following realms:

a) North America   b) Middle America   c) South America
d) the Atlantic Realm e) the Pacific Realm

3.   The Mediterranean climate is classified under which of the following Köppen-Geiger letters:

a) C   b) H   c) D   d) B   e) M

4.   Which of the following realms does not contain a major world-scale population cluster?

a) South America   b) North America   c) Europe
d) South Asia   e) East Asia

5. In its spatial structure, which of the following is likeliest to exhibit a core area?

   a) transition zone   b) regional periphery   c) formal region
   d) uniform region   e) functional region

6. Which of the following is not associated with a region's physiography?

   a) tectonic plate   b) hydrologic cycle c) cultural landscape
   d) effects of glaciation e) dominant vegetation

True-False

1. The world's population is now approximately 6 billion in total size. T

2. Southeast Asia does not rank among the world's three largest population T
   agglomerations.

3. Interglaciations are relatively warm spells that follow glaciations. T

4. Geographic realms almost never change over the course of a millennium. F

5. A city-hinterland relationship is an example of a functional region. T

6. The European state model disappeared after World War II and is no longer evident
   in Europe today. F

Fill-Ins

1. The large country at the center of the realm called South Asia is _India_.

2. The _Functional_ region, also known as the focal region, is marked not by an internal
   sameness but by the integration of a set of places and their interactions.

3. Desert and steppe climates both belong to the Köppen-Geiger category included
   under the letter ___B___.

4. The composite of human imprints on the Earth's surface is called the _cultural landscape_.

5. The two countries that constitute the Austral Realm are Australia and _New Zealand_.

6. A region's _relative_ location refers to its spatial position with respect to
   surrounding regions.

Matching Question on World Realms

B	1.	Contains Far East and Eastern Frontier	A.	Europe
I	2.	Jakota Triangle	B.	Russia
H	3.	Dominated by India	C.	North America
D	4.	Caribbean islands	D.	Middle America
F	5.	The Middle East	E.	South America
A	6.	Contains the British Isles	F.	North Africa/Southwest Asia
G	7.	Contains Equatorial Africa	G.	Subsaharan Africa
KJ	8.	Peninsular mainland, arc of islands	H.	South Asia
L	9.	Melanesia and Polynesia	I.	East Asia
C	10.	Contains the Pacific Hinge	J.	Southeast Asia
E	11.	Strong Spanish/Portuguese imprint	K.	Austral Realm
XK	12.	Insular and separated	L.	Pacific Realm

Essay Questions

1.   Identify and discuss the major geographic characteristics of your home region. Include, in your narrative, substantive answers to the following questions: Is it a formal or functional region, or does it contain aspects of both? How strong a role does urbanism play and, if prominent, are you located in the urban center itself or the hinterland? Are the boundaries of your region sharply defined or do they appear as broad zones of transition?

2.   The cultural landscape of a region is one of its most important geographic components. Define and discuss this concept, taking into account the role of culture itself, its translation into a spatial expression on the land surface, and the significance of the special qualities that offset one cultural landscape from another.

3.   Water has been shown to be one of the most significant variables in the natural environment. Discuss the broad global patterning of precipitation, and the inter-relationship between that distribution and the world pattern of climate.

4.   The spatial distribution of population has been called the most essential of all geographic expressions because it represents the totality of human adjustments to the environment at that moment in time. Identify and discuss the global dis-tribution of humankind, focusing on the four largest population agglomerations and the reasons why so many people have concentrated in those locales.

5.   The *geographic realm* is the regional unit that forms the basis of the study of world geography in this book. Discuss what is meant by this regional term, and using examples chosen from the 12-part scheme introduced on pp. 29-32, list two leading features of four different realms.

TERM PAPER POINTERS

In this section of each chapter in this **Study Guide**, term paper topics will be discussed and some tips provided. Research paper topics can readily be selected from among the major themes emphasized in each chapter of the book, and it should be easy for you to identify several in the Introduction. However, since the aim of this opening chapter is to provide a broad background on world physical and human geography prior to a realm-by-realm survey, it is presumed that your term papers will deal with topics rooted in the coming regional chapters. Nonetheless, you will want to keep in mind the topical coverage of the Introduction chapter because of its worldwide scope as well as the introduction of many concepts, ideas, and terms that are developed further later in the book. For example, if you decided to do a term paper comparing the traditional agricultural systems of India and China, there is much valuable background material here in the Introduction. Among the topics relevant to such a study are environmental differences (apparent in world maps of climate, precipitation, and landscapes) and the economic-geographic properties of densely-populated disadvantaged countries; and among the conceptual themes first introduced here are cultural landscapes, various spatial relationships, geographic realms, and economic development.

One of the most important tasks in preparing a research paper is finding the appropriate literary sources. The **References and Further Readings** section near the back of the textbook has been extensively revised and updated with exactly that in mind, and is well worth consulting after you have chosen your term paper topic. At the same time, it is a good idea to become acquainted with the professional journals of geography that are held by your library. Most of the topics treated in the textbook are covered in these periodicals, which are indexed in the appropriate reference catalogues (and annually in the journals themselves). For the uninitiated researcher, *The Geographical Review*, the *Journal of Geography*, and the *Journal of Cultural Geography* are most convenient to start with. At the more advanced level, there is the *Annals of the Association of American Geographers*, *The Professional Geographer*, the *Geographical Journal*, *Economic Geography*, and *Progress in Geography*; only those with mathematical training, however, should consult *Geographical Analysis*. In larger libraries, you will also find journals on specific subdisciplines (e.g., *The Journal of Historical Geography*, *Urban Geography*, *Physical Geography*) as well as certain world regions (e.g., *Journal of Tropical Geography*, *Chinese Geography* and *Environment*, *Australian Geographer*). In the interdisciplinary arena, geography is frequently treated in *Landscape*, *Environment and Behavior* and *Economic Development and Cultural Change*. Moreover, regional geography topics are widely covered in the regional studies journals that are multidisciplinary in nature; some of the leading ones are the *Journal of Asian Studies*, the *Latin American Research Review*, and the *Journal of Modern African Studies*—do not overlook the journal(s) that covers the part of the world you are studying. Finally, there are several bibliographies available (some are listed in the **References and Further Readings** section). The most significant is *Current Geographical Publications*, published at regular intervals by the American Geographical Society Collection of the University of Wisconsin-Milwaukee Library, an updating of its central catalogue that many university libraries also possess in the form of a series of hardbound volumes in their reference department. Another, still useful guide is *A*

Geographical Bibliography for American Libraries, edited by Chauncy D. Harris, which is listed in the **References and Further Readings** for the Introduction chapter.

You should also be aware of the textbook's Web Site, which includes direct links to a number of other Web Sites that may be quite helpful in your research (http://www.wiley.com/college/regions2000).

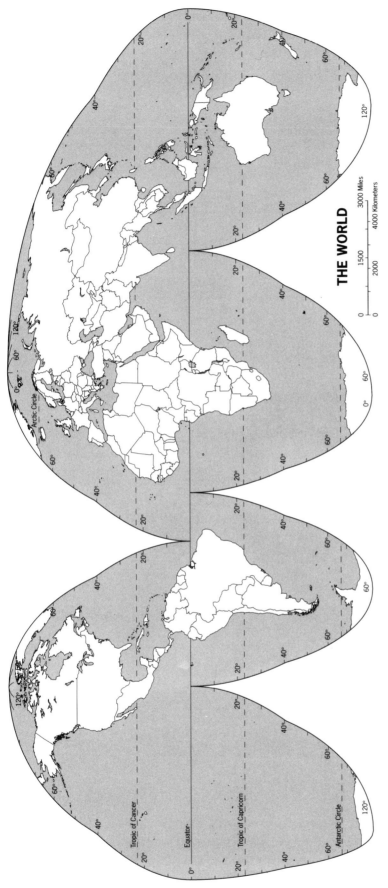

THE WORLD

3000 Miles
4000 Kilometers
1500
2000
0
0

Arctic Circle

Tropic of Cancer

Equator

Tropic of Capricorn

Antarctic Circle

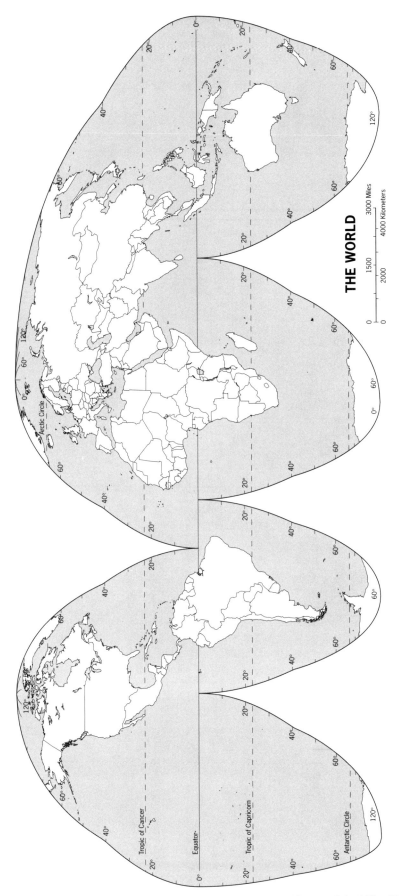

THE WORLD

Arctic Circle

Tropic of Cancer

Equator

Tropic of Capricorn

Antarctic Circle

3000 Miles
4000 Kilometers

1500
2000

0
0

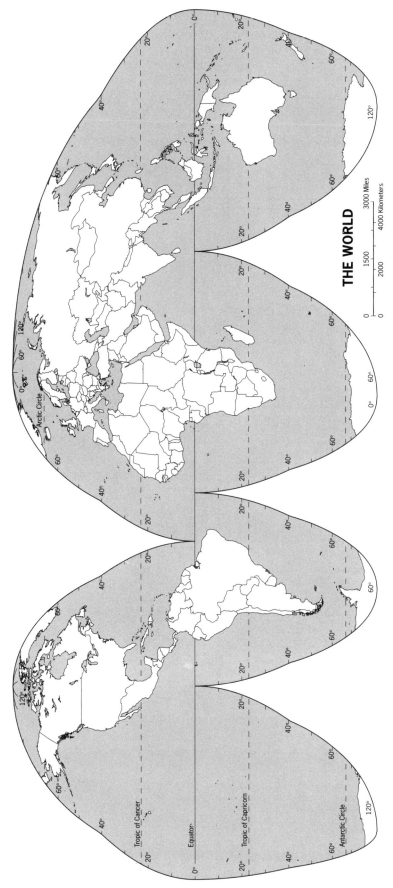

THE WORLD

| 0 | 1500 | 3000 Miles |
| 0 | 2000 | 4000 Kilometers |

Arctic Circle

Tropic of Cancer

Equator

Tropic of Capricorn

Antarctic Circle

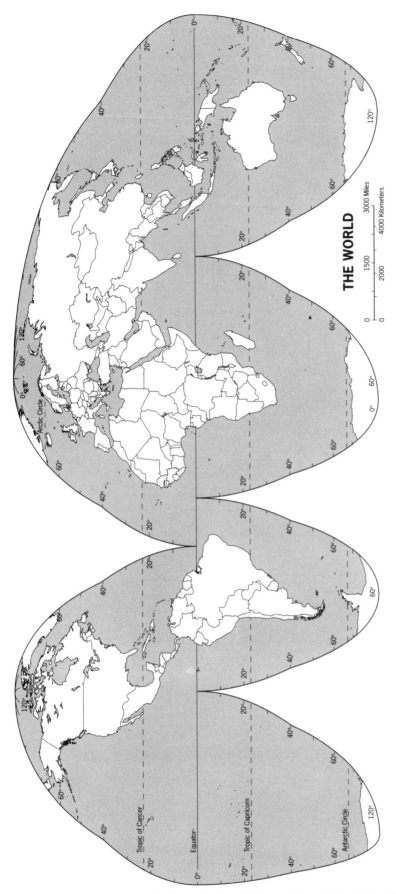

THE WORLD

	1500		3000 Miles
0			
0		2000	4000 Kilometers

Arctic Circle

Tropic of Cancer

Equator

Tropic of Capricorn

Antarctic Circle

CHAPTER 1
EUROPE

OBJECTIVES OF THIS CHAPTER

Chapter 1, which is the first regional chapter, focuses on the geography of Europe. Following an introduction to Europe's advantageous geographic properties, its physical, historical, economic, political, and urban geography are discussed in their realm-wide contexts. The next section treats European supranationalism and considers current trends as well as future prospects for greater unification. The rest of the chapter is devoted to an examination of Europe's internal regions, proceeding through Western Europe, the British Isles, Northern, Mediterranean, and Eastern Europe.

Having learned the regional geography of Europe, you should be able to:

1. Identify Europe's remarkable geographic properties that propelled this modestly-sized corner of the world into international prominence and advanced economic development.

2. Understand Europe's broad physical geography, especially its subdivisioning into four major physiographic regions.

3. Appreciate the rich historical heritage of European society, whose modern foundation rests upon numerous economic and political revolutions.

4. Discern the current geographic dimensions of the realm, including the problems faced by overindustrialized areas in a new postindustrial age.

5. Understand the six-decade-long trend toward greater European unification, the recent progress that has been made, and the challenges faced by the European Union in consolidating its gains and overcoming forces that would drive it apart.

6. Name the five regions of Europe, and the countries they contain.

7. Understand such basic politico-geographical concepts as nation-state, devolution, centripetal/centrifugal forces, irredentism, balkanization, ethnic cleansing, and supra-nationalism.

8. Identify the leading subregions of Western Europe as well as the British Isles, Northern, Mediterranean, and Eastern Europe, and be familiar with the major geographic qualities of each of them.

9. Locate the major features of Europe on a map, including its countries, prominent physical regions, primary rivers, leading industrial areas, and largest urban centers.

GLOSSARY

Continentality *(44)*

An inland location remote from the seas. In this realm, occurring only in easternmost Europe; greater variations in temperature and drier than coastal locales.

Infrastructure *(49)*

The foundations of a society: urban centers, transport networks, communications, energy distribution systems, farms, factories, mines, and such facilities as schools, hospitals, postal services, and police and armed forces.

Lingua franca *(50)*

Refers to a "common language," a second language that can be spoken and understood by many peoples, although they speak other languages at home.

Areal functional specialization *(50)*

The production of a particular good or service as a dominant activity in a particular location.

Renaissance *(50)*

Era in the second half of 15th century Europe, in which national consciousness and pride arose, as well as a renewed interest in Greek and Roman achievements, science, government, and the arts.

Mercantilism *(50-51)*

Protectionist policy of European states during the sixteenth to the eighteenth centuries that promoted a state's economic position in the contest with other countries. The acquisition of gold and silver and the maintenance of a favorable trade balance (more exports than imports) were central to the policy.

Agrarian Revolution *(51)*

The significant change in European farming techniques and equipment that made it possible to sustain large population increases in the 17th and 18th centuries.

Model *(51-box)*

An idealized representation of reality, created in order to demonstrate certain of its properties.

Von Thünen's Isolated State (51-52)

Explains the location of agricultural activities in a commercial, profit-making economy. A process of spatial competition allocates various farming activities into concentric rings around a central marketplace, with profit-earning capability the determining force in how far a crop locates from the market. The original (1826) *Isolated State* model now applies to the continental scale; see Figs. 1-5, p. 52, and Fig. 1-6, p.53.

Industrial Revolution (52-55)

A major turning point in a region's economic development whereby manufacturing becomes a leading growth activity, engendering further technological breakthroughs, capital investments, urbanization, and (eventually) a demographic transition.

Agglomeration (55)

Process involving the clustering or concentrating of people or activities. Often refers to manufacturing plants and businesses that benefit from close proximity because they share skilled-labor pools and technological and financial amenities.

Nation (56)

A group of tightly-knit people possessing bonds of language, ethnicity, religion, and other shared cultural attributes. Such homogeneity actually prevails within very few states.

Nation-state (56)

A political unit wherein the territorial state coincides with the area settled by a certain historical group of people, considering themselves to be a *nation*.

Centripetal forces (56-box)

Forces that tie a state together, unifying and strengthening it.

Centrifugal forces (56-box)

Disunifying or divisive forces that internally weaken a state.

Spatial interaction (58)

Movement of any sort across geographic space, usually involving the contact of people in two or more places for the purposes of exchanging goods or services.

Complementarity (58)

Regional complementarity exists when two regions, through an exchange of commodities, can specifically satisfy each other's demands.

Transferability (58)

The capacity to move a good from one place to another at a bearable cost; the ease with which a commodity may be transported.

Intervening opportunity (58)

The presence of a nearer opportunity or supply source that diminishes the attractiveness of sites farther away.

Primate city (59-60)

A country's largest city—ranking atop the urban hierarchy—most expressive of the national culture and usually (but not always) the capital city as well.

Metropolis (60)

An urban complex consisting of the core or central city and the surrounding ring of suburbs.

Central business district (CBD) (60)

The downtown heart of a central city, the CBD is marked by high land values, a concentration of business and commerce, shopping facilities, and its most prestigious residences.

Devolution (61)

The process whereby regions within a state demand and gain political strength and growing autonomy at the expense of the central government.

Four Motors of Europe (63)

Rhône-Alpes (France), Baden-Württemberg (Germany), Catalonia (Spain), and Lombardy (Italy). Each is a high-technology-driven region marked by exceptional industrial vitality and economic success not only within Europe but on the global scene as well.

Regional state (63)

According to Ohmae, a subnational region with economic power and influence (natural economic zone) that bypasses its traditional national government in its external economic relations, and is shaped by the global economy of which it is a part.

Euroregion (63)

A territorial entity straddling one of Europe's international boundaries, whose purpose it is to foster cooperation and reduce inequalities on each side of that border.

Supranationalism (63-64)

A venture involving three or more national states—political, economic, or cultural cooperation to promote shared objectives. The European Union is such an organization, as is the Organization of Petroleum Exporting Countries (OPEC).

European Union (64-box; 64-66)

Formerly called the Common Market or European Economic Community, it currently consists of 15 member countries: France, Italy, Germany, Belgium, the Netherlands, Luxembourg, the United Kingdom, Ireland, Denmark, Greece, Spain, Portugal, Austria, Finland, and Sweden.

Site (72)

The *internal* locational attributes of a place, including its local spatial organization and physical setting.

Situation (72)

The *external* locational attributes of a place; its *relative location* or position with reference to other non-local places.

Conurbation (76)

General term used to identify large multi-metropolitan complexes formed by the coalescence of two or more major urban areas. Randstad Holland in the Netherlands is a leading European example.

Landlocked location (77)

An interior country that is surrounded by land, such as Switzerland and Austria.

Break-of-bulk point (85)

A location along a transport route where goods must be transferred from one carrier to another. In a port, the cargoes of oceangoing ships are unloaded and put on trains, trucks, or perhaps smaller boats for inland distribution.

Entrepôt (85)

A place, usually a port city, where goods are imported, stored, and transshipped. Thus an entrepôt is a *break-of-bulk point*.

Shatter belt (94)

A zone of chronic political splintering and fracturing—Eastern Europe and Southeast Asia are classic examples.

Balkanization *(94)*

The fragmentation of a region into smaller, often hostile political units.

Ethnic cleansing *(94)*

The forcible ouster of entire populations from their homelands by a stronger power who seeks to conquer their territory.

Exclave *(98)*

A bounded (non-island) piece of territory that is part of a particular state but lies separated from it by the territory of another state. Alaska is an exclave of the United States.

Irredentism *(99-100)*

A policy of cultural extension and potential political expansion by a country aimed at a group of its nationals living in a neighboring country.

SELF-TESTING QUESTIONS

Cover the right side of the page with a sheet of paper. Uncover each line after you have attempted to answer the question in the left column. If necessary, refer to the textbook page(s) listed at the right.

Question	Answer	Page
Europe's Characteristics		
What types of raw materials spawned Europe's development?	Early on rich soils, good fishing waters, domesticatable animals, and plentiful wood for building. Later, mineral fuels and ores made industrialization possible.	44
Has the diversity of ancestries in Europe been an asset or a liability to the realm?	Both— it has generated interaction and exchange, but has also caused conflict and war.	44
What are Europe's chief geographic resources?	Its unmatched advantages of scale and proximity; a diverse environment and cultural mosaic, facilitating innovation and easy contact with the rest of the world.	44-46
Where is Europe's eastern boundary?	Some scholars insist it is the Ural Mountains, while others argue that all of western Russia is a transition zone; this text uses the boundary between Eastern Europe and Russia.	49-box
Physical Landscapes		
What are the major characteristics of the Central Uplands?	Forms a transition zone between the Alps and the North European Lowland; a resource-laden belt where Industrial Revolutions and cities emerged in the 19th century.	46-47
What are the major characteristics of the Alpine Mountains?	The Alps and their outliers (the Pyrenees, Dinaric Alps, and Carpathians) form a high-mountain backbone, separating Mediterranean Europe from the rest of the realm—but have not been a serious obstacle to trade and communication.	47-48
What are the major characteristics of the Western Uplands?	Rugged, older highlands often located along stormy oceanic fringes; represent earlier geo-logical mountain building than the Alpine system.	48

What are the major characteristics of the North European Lowland?	Densest populations of the realm are found here; historic route of human contact and migration, yet much internal variation in generally low-lying terrain; also contains much of Europe's most productive agriculture and a multitude of navigable rivers.	48

Heritage of Order

What area was the heart of the ancient civilization of Greece? And what were some Greek contributions?	The eastern Mediterranean formed the core of a large empire that was eventually defeated by Rome. Political science and philosophy, as well as architecture, sculpture, literature, and education marked early Greek society.	49
What additional contributions did the Romans make to empire-building?	Politico-territorial organization on a much wider scale, extending from Britain to Persia and from the Black Sea to the lower Nile valley; internal diversity was a strength, fostering trade and exchange of ideas and innovations; unparalleled advances in agriculture, political authority, urbanization, transportation, and development of areal functional specialization.	49-50

Decline and Rebirth

What were the lasting contributions of the Romans through the Dark Ages?	Their language and its offshoots, Christianity, educational traditions.	50
What was the overriding politico-geographical trend in Europe between the fall of Rome and the Renaissance?	The fragmentation of large political units into tiny feudal territories; some empires emerged, as did (eventually) elements of the realm's modern political map.	50
What were the main trends of the Renaissance?	Reviving monarchies, early nation-state formation, beginnings of overseas colonial empires, political nationalism, and renewed interest in Greek and Roman achievements.	50
What is *mercantilism*?	The main form of economic nationalism—the acquisition of colonial territories, gold and silver, and favorable international trade—a policy actively promoted by the state.	50-51

The Revolutions

What was the agrarian revolution?	The transformation of European agriculture through changes in land tenure, improvements in farming techniques, equipment, storage, and distribution systems.	51
What is the von Thünen model and what are its lasting virtues?	Johann Heinrich von Thünen in 1826 introduced his model of commercial agricultural spatial organization which still applies to contemporary Europe at the macro-scale (see text pp. 51-52). Basically, agriculture will organize itself into a series of concentric zones around the major urban food market(s), with the most profitable farming activity located in each ring.	51-52
What was the *Industrial Revolution*?	The rapid growth of mass manufacturing through mechanization that triggered far-reaching social and economic change, demographic transition, and mass urbanization.	52-55
Name the factors of industrial location enunciated by Alfred Weber.	"General" factors such as transport costs, and "special" factors such as perishability in the case of certain foods; "regional" factors (transport and labor costs), and "local" factors of *agglomeration* and *deglomeration*; transport costs were seen as highly critical, with industrial plants most attracted to locate at the lowest-transport-cost site(s).	55
How did the political revolution after 1780 forever change Europe?	Monarchies gave way to republics, democracy flourished, and nationalism became the dominant political force.	55-56
What is a *nation*?	A group of tightly-knit people possessing bonds of language, ethnicity, religion, and other shared cultural attributes. Such homogeneity actually prevails within very few states.	56
What is a *nation-state*?	A political unit wherein the territorial state coincides with the area settled by a certain historical group of people, considering themselves to be a nation.	56
What are *centripetal* and *centrifugal* forces?	*Centripetal forces* unify a state; *centrifugal forces* are divisive or disunifying.	56-box

Europe Today

| Is Europe an overall homogeneous regional unit? | To a surprising degree it is not—language, religion, and race are marked by much diversity; yet this diversity, through various subregional complementarities and cooperation, has been overcome to form a well-unified realm. | 56-58 |

| Define the three principles of spatial interaction developed by Edward Ullman. | *Complementarity*, which exists when two places, through an exchange of goods or services, can specifically satisfy each other's demands; *transferability*, the capacity to move an item from one place to another at a bearable cost; and *intervening opportunity*, the presence of a nearer trade opportunity that diminishes the attractiveness of places farther away. | 58 |

| What is the status of Europe's transportation network? | Always highly efficient, it is among the most advanced on Earth; passenger rail service in particular is constantly improving. | 59 |

European Urbanization

| Where does Europe rank among the urbanized realms? | Near the top. No less than 71% of Europe's population live in cities and towns; Western Europe's countries rank in the 80-percent-and-higher range. | 59 |

| What is a primate city? | According to Mark Jefferson's "law," a country's largest city that is simultaneously most expressive of the national culture and often the capital city as well; you should be able to recognize several. | 59-60 |

| How do European cities differ from U.S. cities? | European cities are much older than those of the US, and their layout is more cramped. Available land is scarcer in European cities and European governments tend to take more control in supervising planning than in American cities. | 60 |

| Define *central business district (CBD)*. | A central city's heart, its downtown that is marked by high land values, tall buildings, and the clustering of business, commerce, and government, shopping facilities, and prestigious residences. | 60 |

European Devolution and Unification

What is *devolution*?	The process whereby regions within a state demand and gain political strength and growing autonomy at the expense of the central government.	61
What are the Four Motors of Europe?	1) Rhône-Alpes region, 2) Lombardy, 3) Catalonia, 4) Baden-Württemberg. These subnational regions are hubs of economic power and influence which have developed direct linkages with one another, bypassing the capitals and governments of their countries.	63
What is a *regional state*, and what is a *Euroregion*?	As Ohmae states, a regional state is a subnational region with economic power that in its external relationships bypasses its national government, and deals directly with its global economic partners. A Euroregion is a territorial entity straddling one of Europe's international boundaries, whose purpose it is to foster cooperation and diminish inequalities on either side of the boundary.	63
What are Benelux and the EEC? Name the countries of the European Union.	Benelux (1944) involved the economic association of Belgium, the Netherlands, and Luxembourg; EEC (1958) was the European Economic Community, originally comprised of France, (then West) Germany, Italy, and the Benelux nations. The countries of the European Union today are France, United Kingdom, Germany, Italy, Greece, Denmark, Ireland, Spain, Portugal, the 3 Benelux countries, Finland, Austria, and Sweden.	64-box, 64-66

Western Europe

Which countries constitute this region?	France, Germany, Belgium, the Netherlands, Luxembourg, Switzerland, Austria, and Liechtenstein.	68
How long was Germany partitioned into East and West?	From the end of World War II (1945) until the reunification of Germany in 1990.	68; 70
What geographical features contributed to West Germany's post-WWII "economic miracle?"	Large and varied amounts of raw materials, port access, a location neighboring all other prosperous European nations, efficient surface transport systems, and security protection from NATO all helped West Germany to prosper.	69-70

What is the difference between *site* and *situation*?	*Site* refers to the internal, local physical and other attributes of a place; *situation* is the geographical position of a place or its external locational qualities.	72
How are the economies of Belgium and the Netherlands complementary to each other?	Belgium is a dominantly industrial country producing a large surplus of varied manufactured goods; the Netherlands is a largely agricultural country, producing substantial surpluses of several food commodities.	75-76
Define *conurbation*.	General term denoting a large multi-metropolitan complex formed by the coalescence of two or more major urban areas; Randstad Holland in the western Netherlands is an outstanding example.	76
What is *Randstad-Holland*?	The Dutch "ring-city" *conurbation* linking Amsterdam, The Hague, and Rotterdam.	76
Which leading international organizations are headquartered in Brussels?	Brussels is the co-capital of the European Union and the headquarters of the North Atlantic Treaty Organization (NATO).	76
What are the four languages spoken in Switzerland? Name another Western European country with more than one official language.	German, French, Italian, and Rhaeto-Romansch. Belgium, whose Flemish residents speak a variety of Dutch, and whose Walloon residents speak French.	77
What city and river are synonymous with Austria?	Vienna and the Danube. The river, thanks to pollution, is now iron gray and not blue.	77
In what way does Vienna represent an outpost?	It is the easternmost city of Western Europe, on the doorstep of transforming Eastern Europe.	77

The British Isles

What are the political components of this region?	Britain—containing England, Wales, and Scotland—and Ireland, which contains the Irish Republic (Eire) and Northern Ireland; all but Eire form the United Kingdom. See Fig. 1-17.	77-78
What are the United Kingdom's 5 subregions, and how are they characterized?	Affluent Southern England, struggling Northern England, individualistic Scotland, intractable Wales, and embattled Northern Ireland.	79

Where did England's Industrial Revolution originate and flower?	In the Midlands arc surrounding the southern Pennines (Manchester, Sheffield, Leeds), where coal was plentiful (and in the Northeast too, though to a lesser degree); London was not directly involved.	79
As stagnant Northern England struggles economically, what are the prospects for Southern England?	High-technology and service industries, such as financial, engineering, communications, and energy-related activities are flourishing in the South.	79-81
Which two nearby cities are the industrial and cultural foci of Scotland?	Respectively, Glasgow and Edinburgh.	81
Both Scotland and Wales are potential candidates for *devolution*—define this concept.	Devolution, in political geography, is the process whereby regions within a state demand and gain political strength and growing autonomy at the expense of the central government.	81; 61
Who are the contestants for power in Northern Ireland?	The Protestant majority, who constitute two-thirds of the residents, vs. the 44 percent who are Roman Catholic; the British are trying to prevent a civil war.	82

Nordic Europe (Norden)

Which countries constitute this region?	Norway, Denmark, Sweden, Iceland, Finland, and Estonia.	82
Why is southern Sweden much more densely populated than northern Sweden?	In southern Sweden, relief is lower and more manageable, soils are better, and the climate is milder.	84
How have the seas favored Norway in recent years?	Huge deposits of oil and natural gas in the Norwegian sector of the North Sea yield both fuel supplies and heightened export income.	84
Which Nordic country is a major dairy-product exporter?	Denmark, whose milder climate and undulating terrain are especially favorable.	84-85
What activities form the base of Finland's economy?	Wood and wood products are the major exports. The manufacture of machinery and growth of staple crops also play important roles.	85

Why is Estonia considered to be part of Northern, rather than Eastern Europe?	Estonia shares close linguistic, ethnic, and historical-geographical ties with Finland.	85

Mediterranean Europe

Which countries constitute this region?	Italy, Spain, Portugal, Greece, Cyprus, and Malta.	86
What generalization can be made about the population distributions of this region's countries?	They are decidedly peripheral with large concentrations in productive coastal and riverine lowlands, with more recent growth of major industrial centers in northern Italy and Spain.	87
How do the levels of urbanization and raw materials of this region compare to those to the north?	Urbanization lags considerably, reflecting agricultural bases of the preindustrial era and higher population growth rates. Raw materials are lacking and must be imported from the European core.	87
Which country is Mediterranean Europe's most populous and best connected to the European core?	Italy.	87
Which vital subregion of Italy contains nearly half its population and today functions as its economic core?	The Po River basin of the north, which focuses on Lombardy, containing the largest Italian city (Milan), and also belongs to Europe's core area (Fig. 1-12, p. 67).	88-89
What is meant by the term *Mezzogiorno*?	This is the name for Italy's poverty-stricken and stagnant agricultural south (southeast of the Ancona Line).	88-89
Which countries are located on the Iberian Peninsula?	Spain and Portugal.	90
Where is Spain's major manufacturing zone located?	In Catalonia, in the northeast of the country.	92
How has EU membership affected Greece economically?	There have been economic benefits, however Greece remains the poorest of the EU countries in terms of GNP.	93

How does the northern area of Cyprus differ from its southern counterpart?	The Turkish-proclaimed Turkish Republic of Northern Cyprus is north of the UN "Green Line," contains about 40 percent of Cyprus' territory, and has about 100,000 inhabitants. The south has 60 percent of the territory, about 700,000 inhabitants, and a more prosperous economy with strong links to Europe.	93

Eastern Europe

What is meant by the term *balkanization*? Why is this region so often called a *shatter belt*?	The fragmentation of a region into smaller, often hostile political units. Because of the numerous cultures that have collided here, the resultant shattering of various cultural and political units has produced an especially fragmented ethnic map.	94
What is ethnic cleansing?	The forcible removal of entire populations from their homelands by a stronger power bent on taking their territory.	94
Which countries constitute this region?	Poland, Czech Republic, Slovakia, Hungary, the new Yugoslavia, Bosnia, Croatia, Macedonia, Slovenia, Romania, Bulgaria, Albania, and five former Soviet republics (Latvia, Lithuania, Belarus, Moldova, and Ukraine).	94
Where is Poland's leading industrial area located?	Silesia is located in southwestern Poland, anchored by the cities of Krakow, Katowice, and Wroclaw.	95
What is the significance of Bohemia?	The Czech Republic's core area; the zone around Prague is one of the region's leading manufacturers and a cosmopolitan center with a rich history.	98
Why is the Danube a failure at unifying Eastern Europe?	Apart from the region's political kaleidoscope of rivalries, the river is almost a constant dividing line and did not attract the core areas of many countries it flows through to locate in its valley.	100
Why was Ukraine an economic-geographic cornerstone of the former Soviet Union?	Ukraine is rich in industrial raw materials and contained the most productive environment (soils; climate) for large-scale agriculture in the U.S.S.R.; Ukrainian mineral and energy deposits gave rise to the *Donbas*, until the 1990s, one of Eurasia's leading regions of heavy manufacturing.	100

How was former Yugo-slavia organized when under communist rule?	It was a federation divided up into six internal "republics," based on the Soviet model, each dominated by a major ethnic group.	102
Name the five independent entities of former Yugoslavia.	The new Yugoslavia (Serbia-Montenegro), Bosnia, Croatia, Macedonia, and Slovenia.	102
Why was Slovenia the most successful state to secede from former Yugoslavia?	It is the most remote area from the Serb domain, it is compact and ethnically homogeneous, and had a well developed economy.	102

MAP EXERCISES

Map Comparison

1. Compare the maps of London (Fig. 1-9, p. 60) and Paris (Fig. 1-15, p. 72); what similarities can be observed in their spatial structuring and land-use patterns?

2. Compare the map of Europe's relative world location (Fig. 1-3, p. 46) with the world political map (Fig. I-11, pp. 24-25); discuss the differences between the relative and absolute location of Europe, and how different map projections (see Appendix A) can convey very different perspectives.

3. The boundary of Europe's core area is carefully drawn in Fig. 1-12 (p. 67). Briefly trace that boundary through each country it lies in and, by referring to other maps and appropriate text, justify why the line follows the route it does.

4. The languages of Europe perform a variable function in the realm's political units, sometimes binding people together and sometimes dividing them further. By reviewing Fig. 1-8 (p. 57) and appropriate text, comment on the status of language as a political force in each of the following countries: United Kingdom, Belgium, Switzerland, Poland, Finland, and Romania.

Map Construction (Use outline maps at the end of this chapter)

1. In order to familiarize yourself with Europe's physical geography, place the following on the first of the outline maps:

   a. *Rivers*: Thames, Seine, Loire, Rhône, Garonne, Meuse (Maas), Rhine, Elbe, Danube, Po, Ebro, Tagus, Guadalquivir, Vistula, Oder, Neisse, Tisza, Sava, Dniester, Dnieper

   b. *Water bodies*: Mediterranean Sea, Baltic Sea, North Sea, Irish Sea, English Channel, Straits of Gibraltar, Tyrrhenian Sea, Adriatic Sea, Ionian Sea, Black Sea, Sea of Azov, Bay of Biscay, Gulf of Finland, Gulf of Bothnia

c. *Land bodies*: Ireland, Sicily, Cyprus, Malta, Corsica, Sardinia, Balearic Islands, Shetland Islands, Scandinavian Peninsula, Jutland Peninsula, Iberian Peninsula, Peloponnesus, Crimea Peninsula

d. *Mountains*: Alps, Pyrenees, Appennines, Dinaric Alps, Pennines, Scottish Highlands, Dolomites, Tatras, Ore Mountains (Erzgebirge), Sudeten Mountains, Jura, Cantabrians, Carpathians, Balkans, Transylvanian Alps, Rhodope Mountains, Pindus Mountains

e. *Other landforms*: Meseta, Po Plain, Massif Central, North European Lowland, Bohemian Basin

2. On the second map, political-cultural geographic information should be entered as follows:

   a. Label each country that is listed in Table I-1 table under Europe (text pp. 36-41).

   b. For each of those countries locate and label the capital city with the symbol *.

   c. Reproduce the language map using light pencil coloring (Fig. 1-8, p. 57).

3. On the third outline map, economic-urban information should be entered as follows:

   a. *Europe's regions*: Using a thick line, draw the boundaries of Western Europe, the British Isles, Northern Europe, Mediterranean Europe, and Eastern Europe

   b. *Supranational affiliations*: Color the countries of the European Union green, and countries that do not belong to the European Union yellow.

   c. *Cities* (locate and label with the symbol ●): London, Birmingham, Manchester, Liverpool, Newcastle, York, Leeds, Edinburgh, Glasgow, Dublin, Belfast, Paris, Marseille, Lyon, Bordeaux, Toulouse, Strasbourg, Antwerp, Brussels, Luxembourg, Rotterdam, Amsterdam, Hamburg, Cologne, Essen, Düsseldorf, Frankfurt, Berlin, Stuttgart, Munich, Zürich, Geneva, Vienna, Graz, Copenhagen, Oslo, Bergen, Göteborg, Stockholm, Helsinki, Milan, Turin, Genoa, Venice, Florence, Rome, Naples, Palermo, Madrid, Lisbon, Seville, Barcelona, Athens, Istanbul, Warsaw, Krakow, Katowice, Prague, Bratislava, Budapest, Bucharest, Sofia, Trieste, Belgrade, Ljubljana, Zagreb, Sarajevo, Pristina, Skopje, Tirane, Chisinau, Odesa, Kiev (Kyyiv), Lviv, Dnipropetrovsk, Donetsk, Kharkiv, Mensk (Minsk), Kaliningrad, Vilnius, Riga, and Tallinn

   d. *Economic regions* (identify with circled letter):

      A - the Ruhr
      B - Paris Basin
      C - Rhône-Alpes Region
      D - Baden-Württemberg
      E - Silesia

      F - Randstad Holland
      G - Catalonia
      H - Lombardy
      I - the Donbas

Short-Answer Questions

Multiple-Choice

1. Which of the following mountain ranges is regarded by a number of geographers as Europe's eastern boundary?

   a) Pennines b) Appennines c) Pyrenees d) Urals e) Alps

2. Which of the following cities is located in Italy's core area?

   a) Milan b) Catalonia c) Barcelona d) Naples e) Rome

3. What is Spain's leading industrial area?

   a) Madrid b) Lisbon c) Andalusia d) Catalonia e) Lombardy

4. In which of the following countries did the Industrial Revolution begin?

   a) Russia b) Germany c) Scotland d) England e) France

5. Which of the following countries is a member of the European Union?

   a) Poland b) Norway c) Cyprus d) Austria e) Switzerland

6. Which of the following political entities was not a part of former (pre-1990) Yugoslavia?

   a) Albania b) Montenegro c) Croatia d) Slovenia e) Macedonia

True-False

1. The Republic of Ireland (Eire) is not a part of the state called the United Kingdom. T

2. Although it pursued the acquisition of territory and precious metals, mercantilism was not concerned with actively spreading Christianity throughout the New World. T

3. The spatial interaction principle of *transferability* refers to the capacity to move a good at a bearable cost. T

4. All of the British Isles lie outside of the Central European Upland region. T

5. The term *metropolis* includes both the central city and its suburban ring. T

6. The major internal regions of Italy are called Autonomous Communities. F

Fill-Ins

1. The *Isolated State* model of commercial agricultural spatial organization was devised by _Von Thünen_.

2. The chief territorial occupant of the Iberian Peninsula is the country of _Spain_.

3. A politico-geographical force that operates to unify a country is known as a _centripetal_ force.

4. _Devolution_ is the term used to describe the situation in which the regions or peoples within a state gain political strength and sometimes autonomy at the expense of the center.

5. Greenland is politically affiliated with the Nordic country of _Denmark_.

6. The eastern Mediterranean island of _Cyprus_ is divided between Greece and Turkey.

Matching Question on European Countries

J	1.	Seat of UN World Court	A.	Bosnia ✓
F	2.	Lyon high-tech region	B.	Ukraine ✓
N	3.	Scottish devolution threat	C.	Czech Republic ✓
P	4.	Home of "White" Russians	D.	Denmark ✓
K ✗	5.	Scandinavian Peninsula	E.	Latvia ✓
I	6.	Appennine Mountains	F.	France ✓
B	7.	Crimea Peninsula	G.	Germany ✓
M	8.	Centered on Belgrade	H.	Hungary ✓
O	9.	Catalonian autonomy drive	I.	Italy ✓
D ✗	10.	Jutland Peninsula	J.	The Netherlands ✓
L	11.	Southwestern Iberia	K.	Norway ✓
E	12.	Eastern Europe's northernmost country	L.	Portugal ✓
			M.	Yugoslavia ✓
H	13.	Home of the Magyars	N.	United Kingdom ✓
A	14.	Large Muslim population	O.	Spain ✓
G	15.	Land of *Länder*	P.	Belarus ✓
C	16.	Moravian Gate		

Essay Questions

1. One of the major turning points in European (and world) history was the achievement of the Industrial Revolution. Discuss the sequence of events that marked this economic transformation, highlight its geographic expressions, and show how it helped shape the current map of European manufacturing activity.

2. One of the major geographic qualities of Europe is its strong regional differentiation, which has given rise to a high degree of areal functional specialization affording multiple exchange opportunities. For *three* of the realm's five regions, provide an example of such spatial complementarity and discuss the interaction that it has produced during this century.

3. Review the principles that underlie the von Thünen model, and discuss the various applications of this generalization to Europe in von Thünen's day as well as ours.

4. The partitioning of Germany was one of Europe's major politico-geographical events in this century—a situation that ended in 1990 with the reunification of West and East Germany. Discuss the resource bases and economic-geographic infrastructures of the *Länder* of former East and West Germany, and show how the reunification has produced financial and social strain.

5. Eastern Europe, the theater of cultural collisions and conflicts for centuries, is a classic example of a *shatter belt*. Discuss the concept of the shatter belt and the balkanization it has produced here, and give several examples of persistent cultural fragmentation (and turmoil) in the region today.

TERM PAPER POINTERS

The "Term Paper Pointers" section of the Introduction chapter in this **Study Guide** offered suggestions about approaching research and writing on geographic realms and their components, and you may wish to consult this material if you are undertaking a report on a European region. You should also be aware of the textbook's Web Site (http://www.wiley.com/college/regions2000), which includes direct links to a number of Web Sites that may be quite helpful in your research.

There is a great deal of good, accessible literature on the European realm and many titles are listed in the **References and Further Readings** section near the back of the book. A fine place to start your research, once you and your instructor have agreed on a topic to pursue, is an up-to-date regional geography of Europe. Several good texts exist, and many are of recent vintage. Among them are Berensten, Clout et al., Diem, Jordan, McDonald, Pinder, and Unwin—a particularly useful older book is the one by Gottmann. Several books also focus on individual regions—Baldersheim & Stahlberg, Carter & Maik, Carter & Turnock, Champion, Chinn & Kaiser, Delamaide, Glebe & O'Loughlin, Hall & Danta, "Happy Family?", Hupchick & Cox, John, Johnson, King et al., Kurti & Langman, Matvejevic, Rhodes, Shaw, Turnock (both titles), and Williams (1988); however, books on specific countries are too numerous to be listed among the textbook references. The best approach is to read up on the topic or region in the appropriate work, being careful to spot additional references on your subject; maps are also important, and should be found in these works as well as in specialized atlases on Europe (Brawer, for instance, treats Eastern Europe). You may also find more about your subject in topical geographies of the realm and its components—some examples are Burtenshaw, Coleman, Embleton, Emerson, Fielding & Blotevogel, Hall & White, Heffernan, Hoggart et al., Hudson & Williams, Jensen-Butler, Keating, King, Lever & Bailly, Mazower, O'Dowd & Wilson, Sommers, Spencer, Townsend, and Wintle. And, by all means, do not overlook current-events sources (some examples are Delamaide, Drost, Fernandez-Armesto, Hoffman, Judt, Lewis, Newhouse, and Ohmae); the *New York Times Index* is an especially good place to begin, and periodicals such as *Focus* (published quarterly by the American Geographical Society) and *Geographical Magazine* are other sources to keep in mind. Finally, the trend toward greater international cooperation and unification in Europe is a particularly interesting and important topic, and is covered by Blacksell & Williams, Cole & Cole, Dawson, Murphy, and Williams (1994).

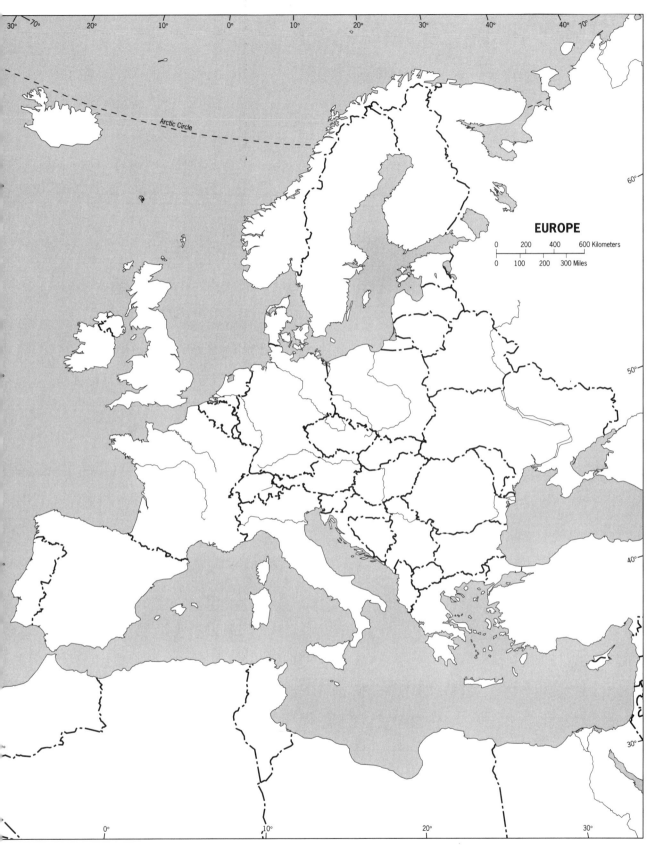

EUROPE

| 0 | 200 | 400 | 600 Kilometers |

| 0 | 100 | 200 | 300 Miles |

Arctic Circle

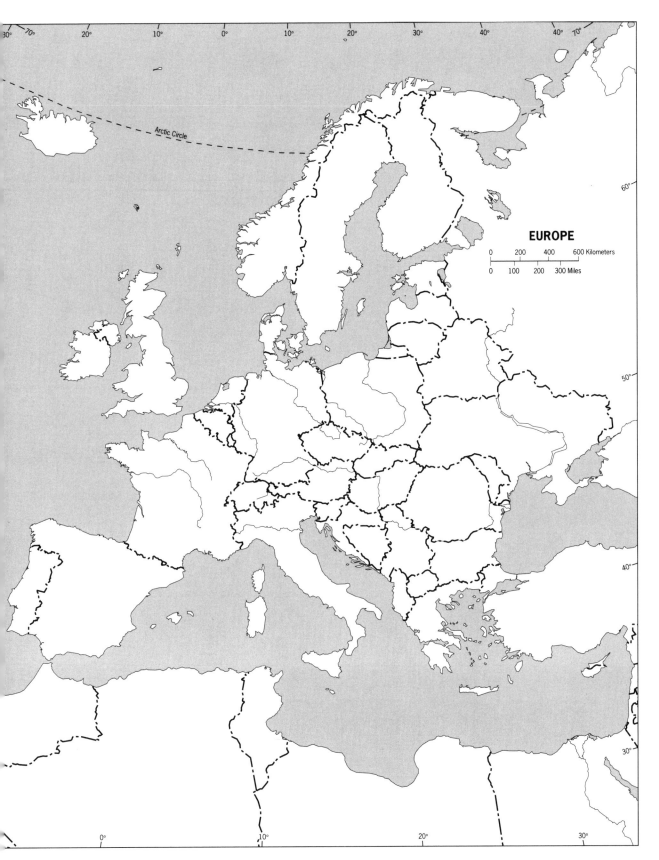

EUROPE

| 0 | 200 | 400 | 600 Kilometers |

| 0 | 100 | 200 | 300 Miles |

Arctic Circle

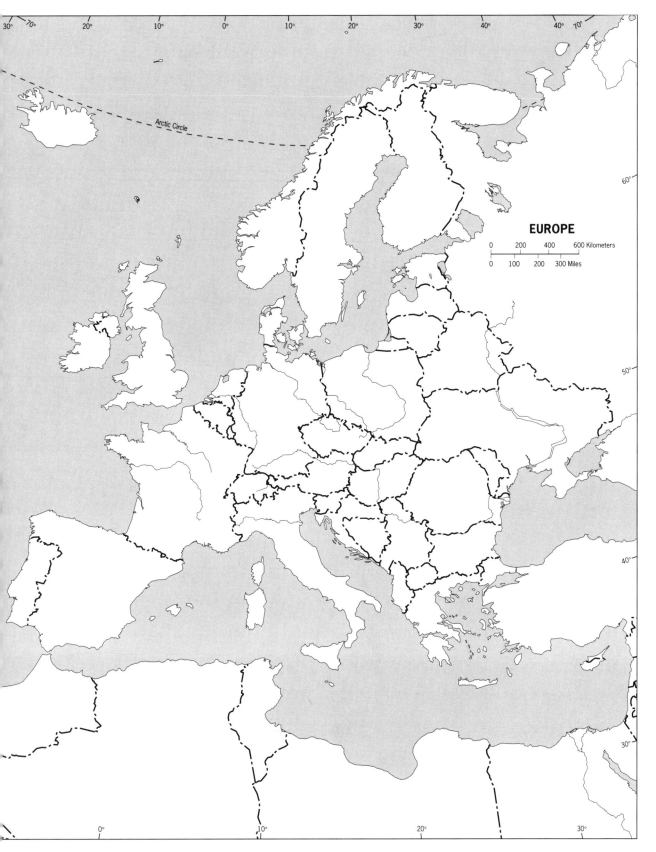

EUROPE

| 0 | 200 | 400 | 600 Kilometers |

| 0 | 100 | 200 | 300 Miles |

Arctic Circle

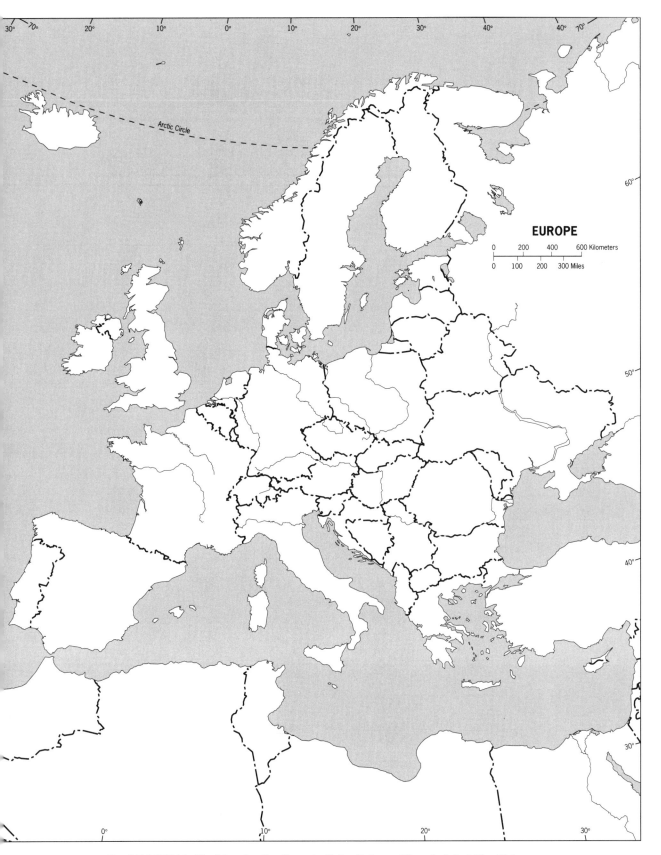

EUROPE

0 200 400 600 Kilometers
0 100 200 300 Miles

Arctic Circle

CHAPTER 2

RUSSIA

OBJECTIVES OF THIS CHAPTER

Chapter 2 covers Russia, former heart of the Soviet Union, a troubled realm, and a society and economy undergoing profound upheaval and transformation. Russia comprises the world's largest political unit in areal size and contains every climate type except the wet tropical category. After the changes engendered by the 1991 collapse of the U.S.S.R. are discussed, Russia's historical cultural evolution is traced. This is followed by reviews of its physiographic, economic, ethnic, and geopolitical frameworks—including a table profiling its 21 internal republics—which set the stage for a survey of the realm's major regions. The chapter also contains an overview of the conflict-plagued Transcaucasian Transition Zone, located in the regional section that covers Russian Peripheries.

Having learned the regional geography of Russia, you should be able to:

1. Understand the overall climatic pattern of this realm.

2. Grasp the essential ingredients of Russia's complex historical evolution, including the contributions of major groups that effected lasting change.

3. Understand the significance of the Soviet period and its legacy for Russia's current circumstances.

4. Understand Russia's cultural-geographic mosaic and its relationship to the churning matrix of internal republics.

5. Appreciate the ongoing spatial reorganization of Russia's economy necessitated by the detachment of its 14 neighboring former Soviet Socialist Republics, especially Ukraine and Kazakhstan.

6. Understand Russia's international political boundary problems, particularly in the eastern reaches of the country.

7. Map and describe the leading functions of the realm's four major regions.

8. Appreciate the ongoing ethnic and territorial conflict in the shatter belt of Transcaucasia.

9. Locate the major physical, cultural, and economic-spatial features of the realm on an outline map.

GLOSSARY

Continentality *(112)*

Refers to degree of inland location away from the moderating effects of the oceans on the climates of adjacent landmasses. The most interior locations, such as eastern Russia, experience both moisture deficits and very large annual temperature ranges.

Climate *(112)*

A term used to convey a generalization or synthesis of all the recorded weather observations over time at a certain place or in a given area; it represents an average of all the weather that occurs there.

Climatology *(112)*

The geographic study of climates. Includes not only the classification of climates and the analysis of their regional distribution, but also broader environmental questions that concern climate change, interrelationships with soil and vegetation, and human-climate interaction.

Weather *(112)*

The state of the atmosphere at a location at any given moment. Recorded in terms of temperature, percentage of humidity, amount of precipitation, wind speed and direction, and the like.

Tundra *(113)*

Treeless plain along the Arctic shore, containing mosses, lichens, and some grasses.

Taiga *(113)*

Mostly coniferous forests that cover large parts of northern Russia and Canada; located south of where the tundra ends.

Permafrost *(113)*

Permanently frozen ground in extreme northern Russia and Canada.

Ostrogs *(116)*

Strategic way-stations built by the Cossacks in river valleys across central and eastern Russia during their 16th and 17th century eastward expansion from Muscovy, in the process defeating the Tatars and consolidating the Russian state all the way to the Pacific coast.

Forward capital (116)

Capital city positioned in actually or potentially contested territory, usually near an international border; it confirms the state's determination to maintain its presence in the region in contention. In the case of St. Petersburg, it demonstrated Russia's commitment to play a major role in the Baltic theater as well as being receptive to European cultural and economic innovations.

Imperialism (117)

The drive toward the creation and expansion of a colonial empire and, once established, its perpetuation.

Soviet Socialist Republic (119-120)

One of the former 15 S.S.R.s that constituted the Soviet Union, each corresponding broadly to one of the country's major nationalities. Largest was the Russian Soviet Federative Socialist Republic (R.S.F.S.R.). Figure 2-5 on p.120 maps the S.S.R.s within the now-defunct Soviet Empire.

Federation (121)

A political framework wherein a central government represents the various subnational entities within a nation-state, where they have common interests--defense, foreign affairs, and the like--yet allows these various entities to retain their own laws, policies, and customs in certain spheres.

Collectivization (121)

The reorganization of a country's agriculture under communism that involves the expropriation of private holdings and their incorporation into relatively large-scale units, which are farmed and administered cooperatively by those who live there.

Sovkhoz (121)

Huge state farm, literally a grain- and/or meat-producing factory, making use of mass labor resources and large-scale agricultural mechanization.

Kolkhoz (122)

A smaller, local collective farm worked by peasants.

Glasnost (121)

Openness.

State Planning (122)

Involves highly centralized control of the national planning process, a hallmark of communist economic systems. Soviet central planners mainly pursued a grand political design in assigning production to particular places; their frequent disregard of economic geography contributed to the eventual collapse of the U.S.S.R.

Unitary state system (128)

Nation-state that has a centralized government and administration that exercises power equally over all parts of the state.

Distance decay (132)

The various degenerative effects of distance on human spatial structures and interactions.

Heartland theory (134)

Geopolitical generalization derived by Sir Halford Mackinder, which saw the center of Eurasian continent as an impregnable fortress against naval power. Mackinder saw the Heartland as the stage for world domination.

Rimland (134)

Counterargument centerpiece in alternative geopolitical scheme to Mackinder's Heartland; concept introduced by Nicholas Spykman, who claimed the rim or outer edge of Eurasia was far more important for potential world domination because of its considerable natural and population resources.

National core area (137)

The heartland of a state. The largest population cluster, the most productive region, the area with greatest centrality and accessibility.

Povolzhye (141)

Russian name for the basin of the middle and lower Volga River.

Transcaucasia (142)

The region constituted by Georgia, Armenia, and Azerbaijan; a mini-shatter-belt known as the "Balkans of Asia;" a historical and current region of ethnic and territorial conflict.

Exclave (142)

A bounded, non-island piece of territory that is part of a particular state, but lies separated from it by the territory of another state. Kaliningrad is an exclave of Russia on the Baltic Sea.

Centrifugal Forces *(144)*

Forces that tend to divide a country—such as internal religious, linguistic, ethnic, or ideological differences.

Kuzbas *(146)*

Contraction of "Kuznetsk Basin," the leading industrial zone of the Eastern Frontier region based on coal, iron ore, and good long-distance transport connections.

SELF-TESTING QUESTIONS

Cover the right side of the page with a sheet of paper. Uncover each line after you have attempted to answer the question in the left column. If necessary, refer to textbook page(s) listed at the right.

Question	Answer	Page
Climatology		
Where is most of Russia's population concentrated?	In "European" Russia, west of the Ural Mountains.	112
What does the field of climatology study?	The distribution of climatic conditions over the Earth's surface and the processes that shape this spatial arrangement.	112
What is the difference between *weather* and *climate*?	Weather is the state of the atmosphere at any given moment; climate is the average weather over a long period of time.	112
Why is farming difficult in Russia?	Even in its most favorable areas, temperature extremes, variable and undependable rainfall, and short growing seasons make agriculture a challenge for Russians.	112
Physiography		
Where is the Russian Plain located?	Across all of "European" Russia as far east as the Urals; the Moscow Basin lies at its heart.	113
What are the two chief rivers of the West Siberian Plain, the world's largest lowland?	The Ob and the Irtysh.	113-114

What is the major inland water body of the Eastern Highlands?	Lake Baykal, the world's deepest.	114
What economic opportunities exist in Russia's harsh north?	Forests provide ample wood for lumbering, a fur trade exists, and there are gold and diamond deposits. Oil and natural gas reserves also are substantial.	114

Russian History

Where was the first Slavic state located?	In present-day Ukraine, northwest of the Black Sea.	114
When did Moscow emerge as the center of the Russian state?	In the 16th century, when Czar Ivan the Terrible began to expel the Tatar conquerors.	115
What was the contribution of the Cossacks to the expansion of modern Russia?	These semi-nomadic peoples drove out the Tatars from Siberia and reached the Pacific by the middle of the 17th century.	116
How far did the Russians penetrate into North America?	They reached across Alaska, the western coast of Canada, and as far south as San Francisco Bay; the U.S. later purchased Alaska from Russia in 1867.	116-box
What were the contributions of Czar Peter the Great and Czarina Catherine the Great to the expansion of the Russian Empire?	Peter opened up the Baltic frontier, built the capital of St. Petersburg, and developed ties with Europe; Catherine pursued the southern frontier, obtaining warm-water ports on the Black Sea.	116

Soviet Legacy

When was the Soviet Union formed?	Between 1917 and 1924, as a result of the Bolshevik Revolution that overthrew the czarist monarchy.	118
Which was the largest of the former S.S.R.s?	The Russian republic, officially known as the Russian Soviet Federative Socialist Republic (R.S.F.S.R.).	120
What was the overall result of past Soviet attempts at population planning?	Minority peoples were generally moved eastward and replaced with Russians. This was termed Russification.	120

Was the former Soviet Union truly a federation?	In theory, yes—any republic wishing to leave the federation supposedly could have. However, in reality, Moscow maintained total control over the republics.	120, 121- box
What were the two main objectives of Soviet economic planners?	To speed industrialization and collectivize agriculture. Soviet planners hoped that efficient farming would free up the labor force necessary for industry.	121
What is the difference between a *kolkhoz* and a *sovkhoz*?	The *kolkhoz* was a local collective farm run by peasants; the *sovkhoz* was a large state farm, mechanized, and run as a food-producing factory.	121-122
Why did the Soviets have to turn to outside sources for food during the decades of communism?	Agriculture encountered enormous difficulties— weak incentives for the farmers, poor management, and constantly recurring weather problems.	122
What was the greatest obstacle to farming in the Kazakh S.S.R.?	The shortage of water, which necessitates irrigation throughout most of the Russian Transition Zone here.	122

Russia Today

What were Mikhail Gorbachev's achievements as a world leader?	Gorbachev opened up the dying Soviet Union to the world once again, removing the Iron Curtain of secrecy and isolation. The collapse of the old order occurred with a minimal loss of life, an amazing achievement.	122-123
When did the Soviet Union collapse?	At the end of 1991.	123
What does the framework of the still-evolving Russian Federation consist of?	The Federation consists of 89 entities; 2 Autonomous Federal Cities, 21 Republics, 11 Autonomous Regions, 49 Provinces, and 6 Territories.	128

Russia's Internal Republics

On the basis of relative location, how are Russia's republics clustered?	The clusters are grouped in the Russian Core, the Caucasus, the Far North, the Central Asian South, and Siberian East.	128
What are the republics of the Russian Core?	Mordoviya, Chuvashiya, Mariy-El, Tatariya, Udmurtiya, and Bashkortostan. See Table 2-2.	129-131

Name the republics of the Caucasian Periphery.	Dagestan, Chechnya, Ingushetiya, North Ossetia, Kabardino-Balkariya, Karachayevo-Cherkessiya, Adygeya, and Kalmykiya. See Table 2-2.	129-131
What are the republics of the Northern Periphery?	Kareliya and Komi. See Table 2-2.	129-131
Name the republics of the Southeastern Periphery.	Altaya, Khakassiya, Buryatiya, and Tuva. See Table 2-2.	129-131
What are the republics of the Eastern Periphery?	Yakutiya. See Table 2-2.	129-131

Political Geography

What is the *heartland theory?*	Mackinder's hypothesis was that whoever controlled the impregnable interior of Eurasia could rule the world.	134
What was the *rimland theory?*	Spykman's counter-hypothesis was that the periphery *(rimland)* of Eurasia contained a greater power potential.	134

Regions of the Russian Realm

List the regions of the Russian realm.	The Russian Core (the Central Industrial Region, the Volga Region, the Urals Region, Southern Peripheries), the Eastern Frontier, Siberia, and the Far East.	137-map; 137
Where is the Russian Core located?	Broadly speaking, the area lying between Russia's western border and the Ural Mountains on the east.	137
Where is the Central Industrial Region?	All definitions are subject to debate, but many geographers define it as the area centering on Moscow, and radiating from it in all directions for about 250 miles.	138
What are Russia's two largest cities?	Moscow is by far the largest, with 9.3 million inhabitants; St. Petersburg (former Leningrad) is the second city, with 5.1 million.	138-140
Which economic sectors have been the focus of development in the Volga *(Povolzhye)* region since World War II?	Great reserves of petroleum and natural gas have been discovered; transportation has also been expanded—the Volga-Don canal now directly connects the Volga waterway to the Black Sea.	141

Transcaucasia

Where is Transcaucasia?	It lies between the Black Sea to the west and the Caspian Sea to the east.	142
Why is Transcaucasia considered to be a zone of political instability?	It is a jigsaw of languages, religions, and ethnic groups—an historical battleground for Christians and Muslims, Russians and Turks, Armenians and Persians.	142
What are the three political entities of Transcaucasia?	Georgia, Armenia, and Azerbaijan, all former S.S.R.s of the Soviet Union.	142

Armenia

Where is Armenia's exclave?	The Nagorno-Karabakh Autonomous Region, inside neighboring Muslim Azerbaijan.	142
Why does Armenia look to Moscow for guidance?	Armenia is potentially vulnerable to Turkish and Azerbaijani enemies, and looks for protection from Moscow because their homeland was secured by Moscow in the Soviet era. The revival of Christianity in Russia also renews linkages, as do mutual agreements over gas pipelines.	142; 144

Georgia

Name the three minority-based autonomous entities in this country.	Abkhazian Autonomous Republic, Adjarian Autonomous Republic, and South Ossetian Autonomous Region.	144
Which centrifugal forces have come to the forefront since the fall of the U.S.S.R.?	Factional fighting destroyed Georgia's first elected government in 1991; conflict continues in South Ossetia as well as in Abkhazia.	144-145

Azerbaijan

What are the republic's ties to its southern neighbor, Iran?	The Shi'ite Muslim Azeris of Azerbaijan have much in common with their ethnic brethren in Northern Iran; irridentism remains a threat.	145
What future political problems does Azerbaijan face?	Fighting over Nagorno-Karabakh with the Armenians, and possibilities of Iranian and Turkish involvement in the dispute.	142; 145

MAP EXERCISES

Map Comparison

1. Russian climates represent one pattern of regional distribution for a large landmass lying between approximately 30°N and 70°N. Compare this pattern to the one exhibited by North America (Fig. I-8, pp. 14-15), and note the similarities and differences.

2. Russia contains a sizeable non-Russian population, as the pre-1992 map (Fig. 2-5, p. 120) reveals. Compare this map with the distribution of Russian ethnic groups (Fig. 2-4, p. 119), and further compare both maps to Russia's population distribution (Fig. I-9, pp. 20-21), noting those areas where the Russians are most under-represented (this should give you an insight into the internal spatial patterns of the 21 republics that are mapped in Fig. 2-7, pp. 126-127).

3. How do the Russian oil and gas regions (Fig. 2-12, p. 141) and railway network (Fig. 2-6, pp.124-125) compare to the national population distribution (Fig. I-9, pp. 20-21)?

4. Compare the maps of Russian (Fig. 2-3, p. 115) and United States settlement expansion (Fig. 3-6, p. 162). What similarities and differences were faced by each country, and did their experiences in the 19th century affect their mutual attitudes toward one another in the 20th?

5. Compare Russia's energy-producing (Fig. 2-12) and manufacturing (Fig. 2-11) regions with appropriate maps in Chapter 1 (covering Ukraine, Moldova, Belarus, Lithuania, Latvia, and Estonia), Chapter 6 (covering the republics of Turkestan: Kazakhstan, Uzbekistan, Turkmenistan, Tajikistan, and Kyrgyzstan), and the section of Chapter 2 covering the Transcaucasus (Georgia, Armenia, and Azerbaijan) that span the 14 S.S.R.s detached from Russia after the 1991 disintegration of the Soviet Union. What are Russia's most important losses? How might these new deficiencies be made up from both inside and outside Russia?

Map Construction

(Use outline maps at the end of this chapter)

1. In order to familiarize yourself with Russian physical geography, place the following on the first outline map:

    a. *Rivers*: Don, Dnieper, Volga, Dvina, Ural, Kama, Ob, Irtysh, Yenisey, Angara, Lena, Aldan, Kolyma, Amur, Ussuri, Tunguska

    b. *Water bodies*: Caspian Sea, Black Sea, Sea of Azov, Lake Baykal, Sea of Okhotsk, Bering Sea, Barents Sea, White Sea, Baltic Sea

c. *Land bodies*: Kola Peninsula, Novaya Zemlya, Kamchatka Peninsula, Kurile Islands, Sakhalin, Moscow Basin, Crimea Peninsula, West Siberian Plain, Yakutsk Basin

d. *Mountain Ranges*: Ural Mountains, Caucasus Mountains, Central Asian Ranges, Verkhoyansk Mountains, Sikhote Alin Mountains

2. On the second map, political-cultural information should be entered as follows:

a. Draw in each internal republic's boundary (see Fig. 2-7, pp. 126-127) and label them all.

b. Reproduce the ethnic map (Fig. 2-4, p. 119) in general form using an appropriate set of area symbols (preferably color pencils).

3. On the third outline map, economic-urban information should be entered as follows:

a. *Cities* (locate and label with the symbol ●): Moscow, St. Petersburg, Rostov, Voronezh, Kursk, Tula, Nizhniy Novgorod, Ivanovo, Yaroslavl, Murmansk, Arkhangelsk, Perm, Kazan, Ufa, Samara, Saratov, Volgograd, Astrakhan, Groznyy, Nizhniy Tagil, Yekaterinburg, Chelyabinsk, Magnitogorsk, Qaraghandy, Astana, Omsk, Novosibirsk, Krasnoyarsk, Bratsk, Irkutsk, Yakutsk, Khabarovsk, Vladivostok, Nakhodka, Komsomolsk, Magadan, Tbilisi, Sokhumi, Batumi, Kutaisi, Yerevan, Gyumri, Baki (Baku)

b. *Economic regions* (identify with circled letter):

A - Central Industrial Region
B - *Povolzhye*
C - Urals Industrial Region
D - *Kuzbas*
E - Far East Region
F - Transcaucasia

PRACTICE EXAMINATION

Short-Answer Questions

Multiple-Choice

1. The political geographer who proposed the *heartland theory* was:

    a) Yeltsin   b) Spykman   c) Köppen
    d) Mackinder   e) Gorbachev

2. What was the capital of the Soviet Union in the early years following the 1917 Bolshevik Revolution?

    a) Moscow   b) Leningrad   c) Petrograd
    d) Stalingrad   e) St. Petersburg

3. Novosibirsk is a leading manufacturing center in the Soviet region called:

    a) European Russia   b) the Far East   c) the *Povolzhye*
    d) the Russian Core   e) the Eastern Frontier

4. Which of the following leaders is a native of Georgia:

    a) Lenin   b) Stalin   c) Yeltsin
    d) Peter the Great   e) Gorbachev

5. The Shi'ite Azeri ethnic group is located in:

    a) the Southern Urals   b) the Russian Far East   c) the St. Petersburg area
    d) Azerbaijan   e) Chechnya

6. The main river of the West Siberian Plain is the:

    a) Amur   b) Volga   c) Ob
    d) Lena   e) Ural

True-False

1. Winters are long, dark, and bitter in most of Russia, but extended, warm summers make for long growing seasons. F

2. Russia is the world's largest state in population size. F

3. In 1992, most of Russia's 90 internal republics would not sign the Russian Federation Treaty, committing to cooperation in a new federal system. F

4. Abkhazia is one of Georgia's internal republics. ✗ T

5. Armenia is landlocked without a coast on either the Black or Caspian Sea. T

6. Sakhalin Island is well endowed with oil and lies just off Russia's Caspian Sea coast. ✗ T

Fill-Ins

1. The Russian word for "openness" is __glasnost__ .

2. The __Volga__ River is most closely associated with the region known as the *Povolzhye.*

3. The U.S. state named __Alaska__ was originally purchased from Russia in 1867.

4. Georgia's political geography is plagued by __centrifugal__ forces.

5. The city formerly called Petrograd and Leningrad is today again called __St. Petersburg__.

6. The leading industrial region of the Eastern Frontier is called the __Kuzbas__ .

Matching Question on Soviet Cities

_D_	1.	*Povolzhye*	A. St. Petersburg ✓
_F_	2.	Naval base on Barents Sea	B. Irkutsk
_A_	3.	Formerly Leningrad	C. Novosibirsk
_G_	4.	Formerly a leading Pacific port in Russo-Japanese trade	D. Volgograd ✓
_H_	5.	Capital of Sakha (the Yakut) Republic	E. Nizhniy Novgorod
_J_	6.	Far East steel center	F. Murmansk ✓
_B_	7.	Lake Baykal	G. Nakhodka
_C_	8.	Kuznetsk Basin	H. Yakutsk
_E_	9.	Soviet Detroit	I. Yerevan
_I_	10.	Armenian capital	J. Khabarovsk

Essay Questions

1. It is claimed that the now-transforming Russian Federation, to a very large extent, is a legacy of St. Petersburg and European Russia—not the product of Moscow and the Communist Revolution. In this context, discuss the course of Russian history and territorial evolution in the two centuries prior to 1917, and conclude your overview with a statement as to whether you concur or not with the above interpretation.

2. The Soviet experiment with a centrally-planned economy was one of the U.S.S.R.'s leading organizational enterprises. Discuss the successes that central planning achieved, its failures, and speculate where the country (assuming it had survived) might today be economically and politically had it chosen a different control structure.

3. It was often said in the now-ended postwar era that communism was an effective system—for the former U.S.S.R. Discuss the overall gains that this kind of economic system brought to the average citizen, its shortcomings, and how the country would change when (and if) a fully capitalist system is put into place.

4. More than most other national capitals and primate cities, Moscow was also the critical focal point for a huge, superpower state. Discuss the historical, cultural, physical, and economic spatial advantages that Moscow possesses that allowed it to take such control of the U.S.S.R. between the early 1920s and the end of 1991. In addition, present a synopsis of the Central Industrial Region that Moscow still anchors, reviewing the productive activities that have concentrated there and further enhanced the geographic position of the capital.

5. Compare and contrast the geographical features of Georgia, Armenia, and Azerbaijan, and evaluate the potential for future development and modernization in each country.

TERM PAPER POINTERS

The "Term Paper Pointers" section of the Introduction chapter in this **Study Guide** offered suggestions about approaching research and writing on geographic realms and their components, and you may wish to consult this material if you are undertaking a report on a Russian region. You should also be aware of the textbook's Web Site (http://www.wiley.com/college/regions2000), which includes direct links to a number of other Web Sites that may be quite helpful in your research.

The disintegration of the Soviet Union, and the emergence of Russia as separate world geographic realm, took place less than eight years prior to the publication of this **Study Guide**. Quite understandably, the literature cannot yet reflect this momentous transformation because it is still going on. Nonetheless, a great deal of the material recently published on the geography of the now-defunct U.S.S.R. is still useful—though you may have to do some updating on your own. Those materials are listed in the **References and Further Readings** section near the back of the book.

Because of Russia's still gargantuan size and complexity, its continuing transformation, and its less-Westernized culture, it is especially important to get hold of a good regional geography that penetrates more deeply into the subject than we are able to in this textbook. Fortunately, several good books are available; among the recommended titles cited in the **References and Further Readings** section are Bater, Dmitrieva, Howe, Lydolph (1990), Shaw (both titles), Symons, and Tomikel & Henderson.. It is also worthwhile to seek additional perspectives. A good source of current maps is Brawer, and Chew contains many historical maps; a compendium of recent information is found in Brown; treatments of the changing Russian scene of the 1990s are provided by Dawisha & Parrott, Diuk & Karatnycky, Dobbs, Dunlop, Kramer, Kraus & Liebowitz, Randolph, Remnick (both titles), "Russia Reborn," Silverman & Yanowitch, and Smith. Historical surveys can be found in Chew, Dunlop, Harris (both titles), Hauner, Hosking, Mastyugina & Perepelkin, and Thompson. Increasingly important ethnic and social considerations are treated in Chinn & Kaiser, Demko et al., Duncan & Holman, Forsyth, Kaiser, and Wixman.

Several *topical* treatments of Soviet geography also exist (Bater, Shaw [1995], and Symons are the most recent overviews). The resource base is covered in Jensen et al. and Pryde. Regional development is highlighted in "A Caspian Gamble," Akaha, Dmitrieva, Gachechiladze, Goltz, Hauner, Horensma, Hunter, Huttenbach, Ioffe & Nefedova, Lincoln, Mote, Rodgers, and Valencia. Political geography topics are treated in Chinn & Kaiser, Dawisha & Parrott, Diuk & Karatnycky, Dunlop, Gachechiladze, Hauner, Hunter, Kaiser, Kraus & Liebowitz, Mackinder, Nijman, Remnick (both titles), Shlapentokh & Yanowitch, and Thompson. The monumental environmental problems of the realm are treated in Lyndolph (1977), Peterson, Pryde, "The Rape of Siberia," and Stewart.

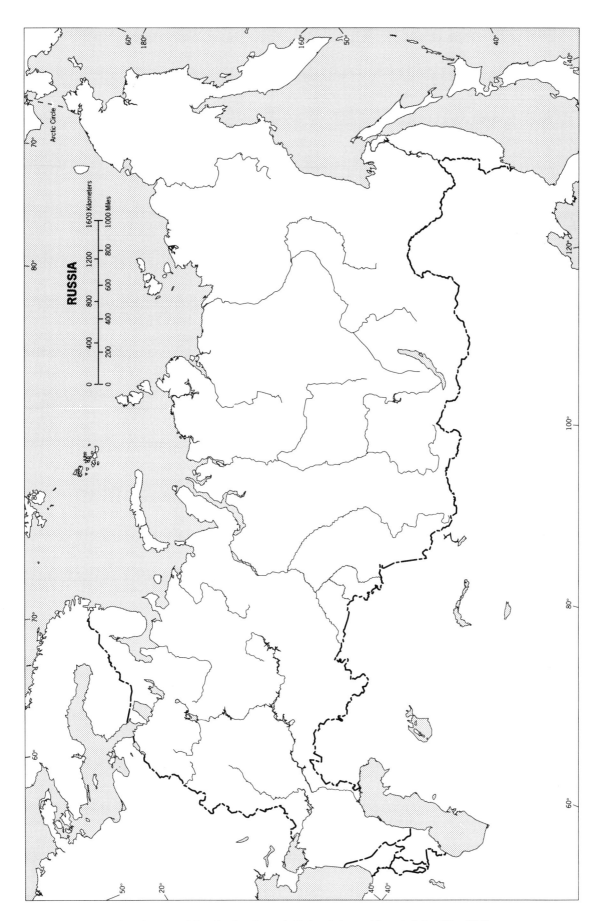

RUSSIA

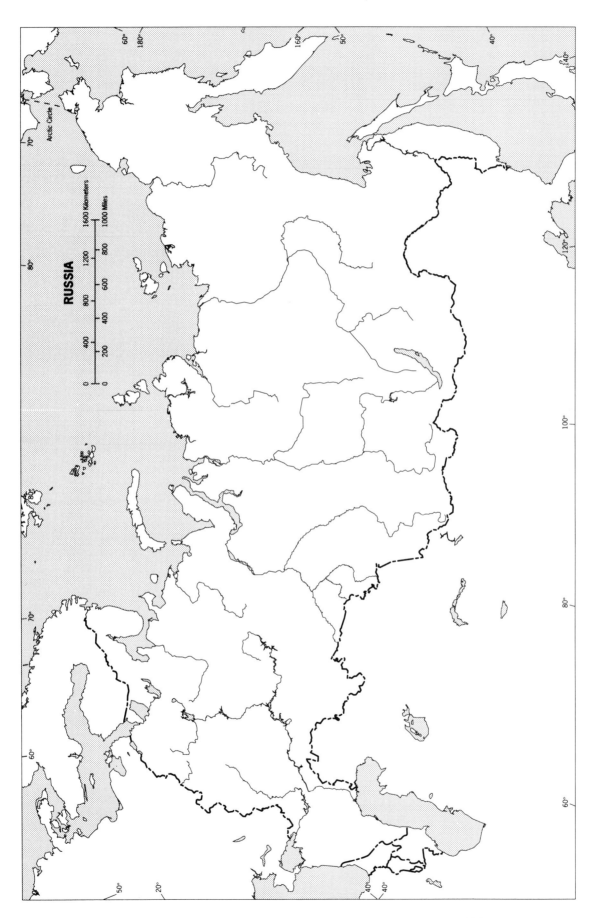

RUSSIA

Arctic Circle

1600 Kilometers
1000 Miles
1200
800
600
400
200
0

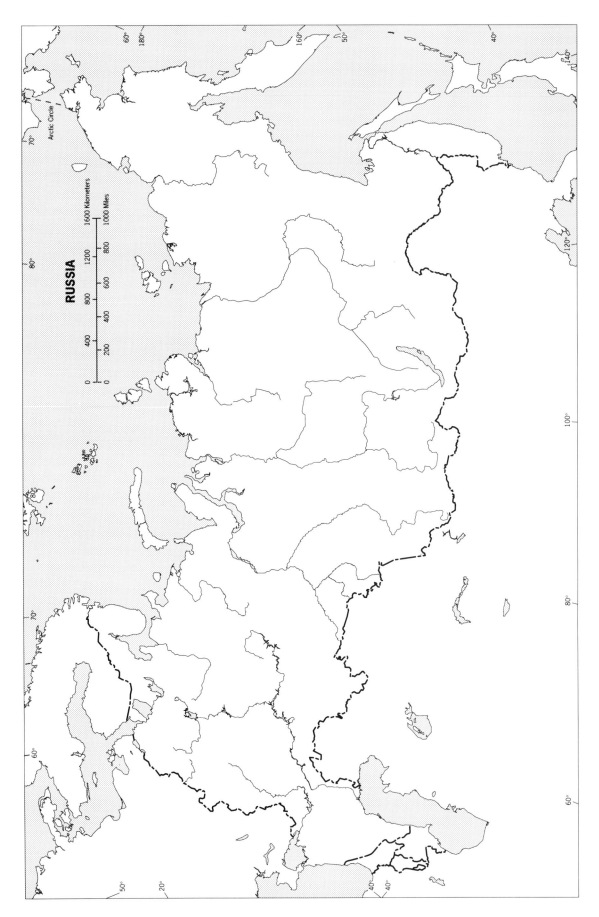

RUSSIA

Arctic Circle

1600 Kilometers
1000 Miles
1200
800
600
400
200
0

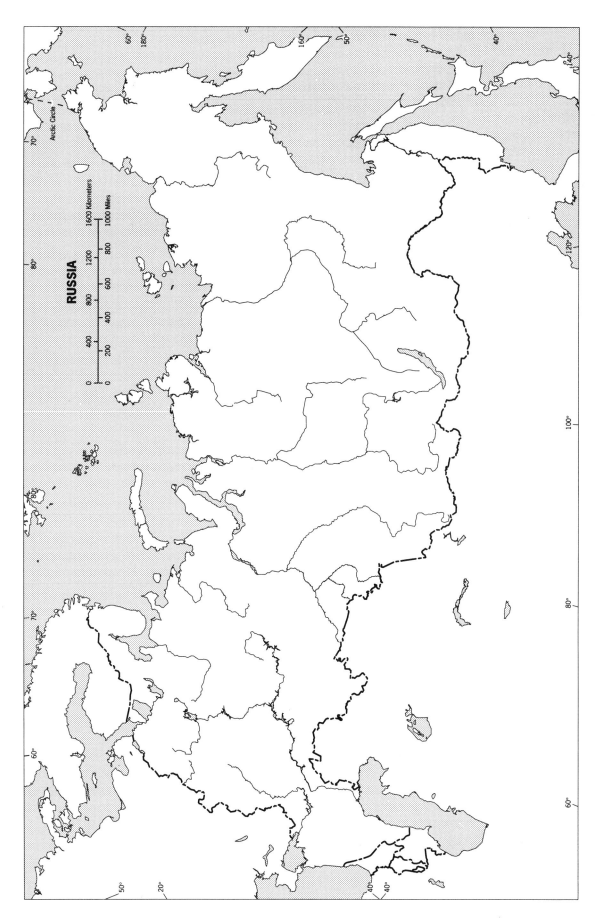

RUSSIA

Arctic Circle

1600 Kilometers
1000 Miles
1200
800
600
400
200
0

CHAPTER 3
NORTH AMERICA

OBJECTIVES OF THIS CHAPTER

Chapter 3 covers the part of the world that most of you know best, but rapid change persists in the United States and Canada as the new century opens. As North America's societies and economies complete their postindustrial transformation, a new geographic reality is emerging. Although this transition is not yet complete, many of its broad features have become evident since 1980 and they form the foundation for this re-examination of the realm's regional human geography. The new dominance of urban realms and the outer suburban city on the metropolitan scene, the economic-geographic significance of the Silicon Valleys and other lineaments of postindustrial society, the continuing demographic shifts reported by the latest census—these are just some of the developments that are reshaping North America which are of such recent vintage that they were still years in the future when the first edition of the textbook appeared in 1971.

After the background introduction to the contemporary United States and Canada—which focuses on the unique composition of its populations—the continent's physical environment is reviewed. The central part of the chapter separately treats the U.S. and Canada, but follows the identical topical sequence. Population distribution is surveyed in historical spatial context, emphasizing settlement patterns in both rural and urban America; the latter metropolitan dimension is examined at some length, covering both macro-scale and internal urban-growth patterns into the new millennium. Cultural geography follows, and then we turn to economic geography, stressing the emerging postindustrial sectors and re-examining agriculture and manufacturing from this new perspective. The concluding regional survey integrates most of the chapter's themes.

Having learned the regional geography of the United States and Canada, you should be able to:

1. Understand the similarities and differences of the U.S. and Canadian populations.

2. Grasp the essentials of the physical geography of this realm.

3. Trace the historical geography of settlement across rural America in the 19th century, and within urban America since 1900.

4. Understand the main processes and patterns of metropolitan growth, at both the scales of the national urban hierarchy and the internal structuring of the individual metropolis.

5. Grasp the emerging patterns as postindustrialism transforms culture and the economy, and urban and regional spatial change intensify.

6. Understand contemporary American economic geography—the shifting resource base, energy patterns, the historical Thünian framework of agricultural regionalization, the dilemmas of declining "smokestack" industry, and the postindustrial revolution's far-reaching spatial impacts.

7. Understand the forces that shaped contemporary Canada—and may split it asunder as Quebec ponders independence.

8. Understand the changing regional infrastructure of North America.

9. Locate the major physical, cultural, and economic-spatial features of the realm on an outline map.

GLOSSARY

Fragmented state *(155)*

A state whose territory consists of several separated parts, not a contiguous whole.

Francophone *(155)*

Country or region in which several languages are spoken, but where French is the *lingua franca* or language of the elite; Quebec constitutes Francophone Canada.

Cultural pluralism *(155-156)*

A society composed of multiple social groups in a single state that do not mix. In Canada, cultural divisions run along ethnic and linguistic lines; in the U.S., the major social cleavage has occurred along racial and economic lines, which still fosters widespread residential segregation.

Residential segregation *(156)*

The refusal of whites to share their immediate living space with nonwhites across most of the U.S. results in the concentration of whites and blacks into separate, racially-distinct neighborhoods.

Physiographic province *(156)*

The clear and well-defined divisioning of a continental-scale land surface into physically uniform regions called physiographic provinces; each is marked by a general homogeneity in relief, climate, vegetation, soils, and other environmental variables.

Rain shadow effect *(158)*

The relative dryness in areas downwind of, or beyond, mountain ranges caused by *orographic precipitation*, wherein moist air masses are forced to deposit most of their water content.

Orographic precipitation *(158)*

Mountain-induced precipitation, especially where air masses are forced over topographic barriers. Areas beyond such a mountain range experience the *rain shadow effect*.

Isohyet *(158)*

A line connecting all places receiving the same amount of annual precipitation, such as the 20-inch (50-cm) isohyet that divides Arid from Humid America (see below). An *isoline*, of which an isohyet is one specific example, is a line connecting all points receiving the same value of a phenomenon (temperature, elevation, air pressure, and the like).

Arid and Humid America *(158)*

The broad divisioning of natural environments in the United States, with the wide boundary following, more or less, the 20-inch (50-cm) *isohyet*—the line along which that exact amount of precipitation is received annually. The average position of this north-south line is approximately 100°W longitude, located in the central Great Plains; to the east is Humid America and to the west lies Arid America (except for the narrow moist zone between the Pacific coast and its nearby parallel ranges). Humid America is generally associated with natural forest vegetation and acidic soils, whereas Arid America contains steppe or short-grass prairie vegetation and alkalinic soils.

Migration *(161-box)*

A change in residence intended to be permanent.

Push-pull concept *(161-box)*

The idea that migration flows are simultaneously stimulated by conditions in the source area, which tend to drive people away, and by the perceived attractiveness of the destination.

Sunbelt *(161)*

The popular name given to the southern tier of the United States, which is anchored by the mega-States of California, Texas, and Florida. Its warmer climate, superior recreational opportunities, and other amenities have been attracting large numbers of relocating people and activities since the 1960s; broader definitions of the Sunbelt also include much of the western U.S., particularly Colorado and the coastal Pacific Northwest.

Culture hearth (163)

Heartland, source area, innovation center, place of origin of a major culture.

Borchert model of United States urban-system evolution (163-164)

A generalization of the historical growth of the U.S. urban system according to key changes in transportation and industrial energy-use that occurred within five stages of development: the Sail-Wagon Epoch (1790-1830), the Iron Horse Epoch (1830-1870), the Steel-Rail Epoch (1870-1920), the Auto-Air-Amenity Epoch (1920-1970), and the Satellite-Electronic-Jet Propulsion Epoch (1970-present).

Core area (Continental Core Region) of North America (164)

As mapped in Fig. 3-7 (p. 164), the rectangular *American Manufacturing Belt* of the northern/eastern U.S. and southeastern Canada—cornered by Boston, Milwaukee, St. Louis, and Baltimore—and containing the industrial districts listed on p. 164.

Megalopolis (164)

When spelled with a lower-case m, a synonym for *conurbation*, one of the large coalescing supercities forming in diverse parts of the world. When capitalized, refers specifically to the multi-metropolitan corridor that extends along the northeastern U.S. seaboard from north of Boston to south of Washington, D.C. Includes Boston, New York, Philadelphia, Baltimore, and Washington, D.C.

Main Street (164)

Canada's dominant megalopolis, the country's primary conurbation stretching southwest from Quebec City through Montreal and Toronto to Windsor on the U.S. border across the river from Detroit.

Adams model of intraurban structural evolution (165-166)

A generalization of the historical growth of the American metropolitan city according to key internal transportation breakthroughs that occurred within four stages of development: the Walking-Horsecar Era (pre-1888), the Electric Streetcar Era (1888-1920), the Recreational Automobile Era (1920-1945), and the Freeway Era (1945-present). See Fig. 3-9 (p. 166).

Ghetto (166)

An intraurban region marked by a particular ethnic character. Often an inner-city poverty zone, such as the black ghetto in the American central city. Ghetto residents are involuntarily segregated from other income and racial groups.

Outer City (166-167)

The outer metropolitan ring, no longer "sub" to the "urb," which is now the essence of the contemporary U.S. city as it acquires a critical mass of population and activities and becomes co-equal to the central city that spawned it.

Suburban downtown (167)

Significant concentration of diversified economic activities centered around a highly accessible suburban location—usually focused on very large regional shopping centers—that are the new automobile-age equivalent of the central-city downtown (CBD).

Urban realms model (167)

A spatial generalization for the large, contemporary U.S. city. It is shown to be a widely dispersed, multi-centered metropolis consisting of increasingly independent zones or realms. See Figs. 3-10 and 3-11 (p. 167).

Dialect (168)

Regional or local variation in the use of a major language, such as the distinctive accents of many residents of the U.S. South or rural New England.

Mosaic culture (169)

The ongoing fragmentation of U.S. social groups into smaller and more specialized communities, stratified not only by race and income but also by age, occupational status, and especially lifestyle.

Primary economic activity (170)

Activities engaged in the direct extraction of natural resources from the environment—particularly mining and *agriculture*.

Secondary economic activity (170)

Activities that process raw materials and transform them into finished industrial products; the *manufacturing* sector.

Tertiary economic activity (170)

Activities that engage in *services*—such as transportation, banking, retailing, finance, education, and routine office-based jobs.

Quaternary economic activity (170)

Activities engaged in the collection, processing, and manipulation of *information*.

Quinary economic activity *(170)*

Managerial or control-function activity associated with *decision-making* in large organizations.

Fossil fuels *(170)*

Energy resources formed by the geologic compression and transformation of ancient plant and animal organisms—*coal, petroleum (oil),* and *natural gas*—which still account for an overwhelmingly large proportion of U.S. energy consumption.

Economies of scale *(173)*

The savings that accrue from large-scale production whereby the unit cost of manufacturing decreases as the level of operation enlarges.

Historical inertia *(173)*

The need to continue utilizing expensive manufacturing equipment and other capital facilities for their full lifetimes in order to cover initial long-term investments; a major reason for the persistence of the American Manufacturing Belt, despite the obsolescence of much of its physical plant.

Postindustrial economy *(174)*

Emerging economy in the United States, Canada, and a handful of other highly advanced countries as industry gives way to a high-technology productive complex dominated by services, information-related, and managerial activities.

Technopole *(174)*

A planned techno-industrial complex (such as California's Silicon Valley) that innovates, promotes, and manufactures the products of the postindustrial informational economy.

Ecumene *(176)*

The habitable portions of the Earth's surface where permanent human settlements have arisen.

Yeates model of Canadian urban-system evolution *(177)*

A generalization of the historical growth of the Canadian urban system according to key changes in transportation and industrial energy-use that occurred within three stages of development: the Frontier-Staples Era (pre-1935), the Era of Industrial Capitalism (1935-1975), and the Era of Global Capitalism (1975-present).

Regional state (180)

A "natural economic zone" that defies political boundaries, and is shaped by the global economy of which it is a part; its leaders deal directly with foreign partners and negotiate the best terms they can with the national governments under which they operate.

North American Free Trade Agreement (NAFTA) (180; 181-box)

Economic alliance that took effect in 1994, steadily eliminating trade barriers between the United States, Canada, and Mexico for agricultural and manufactured goods, as well as tertiary and quaternary services.

Pacific Rim (194-box)

A far-flung group of countries and parts of countries (extending clockwise on the map from New Zealand to Chile) sharing the following criteria: they face the Pacific Ocean; they evince relatively high levels of economic development, industrialization, and urbanization; their imports and exports mainly move across Pacific waters.

SELF-TESTING QUESTIONS

Cover the right side of the page with a sheet of paper. Uncover each line after you have attempted to answer the question in the left column. If necessary, refer to the textbook page(s) listed at the right.

Question	Answer	Page

Two Highly Advanced Societies

What is a *postindustrial* economy?	One dominated by the production and manipulation of information, skilled services, and high-technology activities.	154
What are some major geographic differences between the U.S. and Canada?	The U.S. has a much bigger population, but a smaller area; the U.S. population is widely dispersed, whereas the Canadian people are highly concentrated along their southern border.	154-155

North American Characteristics

What percentage of the Canadians speak English as a primary or home language? French?	English is the home language of 60% of Canada's citizens; 24% speak French as their primary language.	155

What are Canada's two largest provinces in population size?	Ontario and Quebec.	155
Where is the heart of Francophone Canada?	Quebec, where over 85 percent of the people are French Canadian.	155
What are the different leading social divisions in the U.S. and Canada?	In Canada, along linguistic and ethnic lines; in the U.S., by race and income.	155-156

Physical Geography

What is a physiographic province?	A physically-uniform region in terms of topography, climate, vegetation, soils, and other environmental variables.	156
What is the *rain shadow effect*?	The relative dryness in areas downwind of mountain ranges, caused by *orographic* precipitation, wherein moist air masses are forced to deposit most of their water content in the highlands.	158
What are Arid and Humid America?	The dry and moist halves of the conterminous U.S., roughly divided by the transition zone along the 20-inch (50-cm) *isohyet*.	158
What are the broadest vegetation divisions in the U.S. environment?	Arid America contains grassland vegetation; Humid America contains forest vegetation.	158
Name the five Great Lakes and their outlet to the sea.	Lakes Superior, Michigan, Huron, Erie, Ontario; the St. Lawrence River.	158
Name the major tributaries of the Mississippi-Missouri river network.	The Ohio, Tennessee, and Arkansas rivers.	158

Population and Urbanization

What is currently the most significant internal migration trend in the United States?	The persistent drift of people and livelihoods towards the South and West (the "Sunbelt").	160-161

What were the three separate *culture hearths* of the United States?	New England, the Middle Atlantic area, and Tidewater Maryland and Virginia.	162-163
When did the Industrial Revolution occur in the U.S.?	It began in the 1870s, and within 50 years America was the leading industrial power of that time.	163
Name the 5 stages in Borchert's model of American metropolitan evolution.	The Sail-Wagon Epoch (1790-1830), the Iron Horse Epoch (1830-1870), the Steel-Rail Epoch (1870-1920), the Auto-Air-Amenity Epoch (1920-1970), and the Satellite-Electronic-Jet Propulsion Epoch (1970-present).	163
What is the *core area* of the realm, and what are the four corners of the rectangular boundary that encloses this region?	The American Manufacturing Belt, cornered by the cities of St. Louis, Milwaukee, Boston, and Baltimore.	164
What is the economic heartland of this core region?	The Atlantic Seaboard Megalopolis (its Canadian equivalent is *Main Street*, stretching from Quebec City southwest to Windsor).	164
Name the 4 stages in the Adams model of intra-urban structural evolution.	The Walking-Horsecar Era (pre-1888), the Electric Streetcar Era (1888-1920), the Recreational Auto Era (1920-1945), and the Freeway Era (1945-present).	165
What is meant by the terms *outer city, suburban downtowns,* and the *urban realms model*?	The newly urbanized suburban ring, now co-equal to the central city that spawned it; suburban downtowns anchor the outer ring of the multi-centered metropolis; the urban realms generalization views today's metropolis as consisting of several self-sufficient sectors, each organized around its own downtown.	166-167

Cultural Geography

What are the values that underlie the culture of the United States?	Love of newness, desire to be near nature, freedom to move, individualism, societal acceptance, aggressive pursuit of upward mobility, and a firm sense of destiny.	168
What is a *dialect*?	A local variation in a language spoken across a large area.	168

What is the emerging *mosaic culture*?	The ongoing fragmentation of American society into a plethora of narrowly-defined communities, not only along income, racial, and ethnic lines, but also according to age and lifestyle.	169

Economic Geography

Identify: (1) primary, (2) secondary, (3) tertiary, (4) quaternary, and (5) quinary economic activity.	(1) The extractive sector, especially mining and agriculture; (2) manufacturing; (3) the services sector; (4) the information sector; (5) managerial decision-making in large organizations.	170
Name the three *fossil fuels* and their three leading North American source areas.	*Coal*—Appalachia, the northern Great Plains, Midcontinent; *petroleum* (oil)—Gulf Coast, Mid-continent, and Alaska; *natural gas*—Gulf Coast, Midcontinent, and Appalachia.	170
Describe the application of the von Thünen Model to the explanation of the spatial structure of U.S. agriculture.	Like modern Europe, the model fits at the macro-scale, with agricultural regions of decreasing intensity concentrically arranged with increasing distance from the central "supercity"—the northeastern Megalopolis.	172-173
What are *economies of scale*?	In industrial location, the savings that accrue from large-scale production wherein the cost of manu-facturing a single item decreases as the level of the operation enlarges.	173
How does the principle of *historical inertia* help to maintain the aging industries of the Manufacturing Belt?	Because of the need to continue using expensive manufacturing facilities for their full lifetimes to cover long-term investments that were made decades ago.	173
Where has the industrial revitalization effort been centered in the U.S.?	Much of the activity to create high-tech "factories of the future" has been in the Midwest portion of the Manufacturing Belt.	173-174
What are the growth industries in the post-industrial society of the U.S.?	High-technology, white-collar, office-based activities.	174

What are the location factors that attract today's pacesetting "high-tech" companies?	A major university, highly-skilled labor, close proximity to a cosmopolitan urban area, abundant venture capital, an economic climate that supports risk-taking, a locally based network of global business linkages, and a high amenity environment with good housing, pleasant weather, and recreational opportunities.	174
What is a *technopole*?	A planned techno-industrial complex (like California's Silicon Valley) that innovates, promotes, and manufactures the products of the postindustrial informational economy.	174

Canada's Human Geography

Name Canada's Atlantic Provinces.	Nova Scotia, Prince Edward Island, New Brunswick, and Newfoundland.	175
Which two Canadian provinces contain the largest populations?	Ontario and Quebec.	175
When did Canada become a federation?	In 1867, under the British North America Act.	176
Have Canadian metropolises deconcentrated as much as their American counterparts?	Until 1990, they had not–however, today Canada's large cities are becoming increasingly suburbanized: new suburban downtowns anchor ultramodern business complexes.	177
How has the separatist movement fared in Quebec since the 1970s?	English domination of Quebec's culture has been eliminated; a 1980 referendum rejected secession from Canada, but the idea has been revived and the 1995 referendum almost resulted in the approval of secession.	178
What problems afflict Canada's industrial areas?	Postindustrialism has caused massive underemployment in the manufacturing sector because new high-tech industries are not especially labor intensive.	179
What percentage of Canada's total workforce is employed in the tertiary and quaternary economic sectors?	Close to 75 percent.	179

North American Regions

What is the Continental Core?	The Manufacturing Belt, now facing problems in keeping abreast of the realmwide transition from industrial to postindustrial power.	181-182
What problems do northern New England and Atlantic Canada share?	Both are generally rural, possess difficult environments, and were historically bypassed in favor of more dynamic and fertile inland areas; although agriculture and fishing are no longer growth industries, tourism presents an opportunity in this scenic region.	185
What are the South's persistent economic problems?	Uneven development has favored certain areas, left others untouched by recent progress; much growth has occurred at the edges of the region; and Southern central cities increasingly face the problems of their Northern counterparts while their suburbs thrive.	188-189
What are the tricultural influences evident in the Southwest?	The growing Anglo influence, the persistently Hispanic flavor of local cultures, and the sporadic Native American presence.	189
Where is the hottest growth area of the Western Frontier?	The fastest-growing metropolis in the U.S. is the Las Vegas Valley, with its boom triggered by Las Vegas' recreation industry, and economic development fueled by the influx of high-tech and professional service firms.	191
What does the term Pacific Hinge refer to?	The U.S. region that forms an interface between North America and the Pacific Rim; it is the North American gateway to opportunities blossoming on the distant shores of the Pacific Basin.	194

MAP EXERCISES

Map Comparison

1. The Canadian population has always adhered closely to the country's southern border. By comparing the maps of Canada (Fig. 3-15, p. 175), 2000 population distribution (Fig. 3-4, p. 160), and European settlement patterns (Fig. 3-6, p. 162), record your observations in detail about this spatial pattern over the past two centuries.

2. Compare the map of North American physiography (Fig. 3-1, p. 152) and European settlement patterns (Fig. 3-6, p. 162). What relationships are apparent in attraction and avoidance between landform regions and population patterns? Differentiate between 19th and 20th century settlement patterns, and single out those areas where a reversal of settlement preferences seems to have occurred after 1900.

3. Study the map of North American Manufacturing (Fig. 3-7, p. 164), and offer observations about the concentration of industrial activity in the American Manufacturing Belt with respect to other parts of the United States.

4. Study the map of North America's leading deposits of fossil fuels (Fig. 3-13, p. 171). Learn the distributions of each energy resource by listing the major regions where coal, oil, and natural gas, respectively, are produced.

Map Construction   *(Use outline maps at the end of this chapter)*

1. In order to become more familiar with the North American environment, place the following physical-geographic information on the first outline map:

    a. *Rivers*: Connecticut, Hudson, Delaware, Susquehanna, Potomac, Ohio, Tennessee, Mississippi, Missouri, Arkansas, Platte, Red River of the North, Rio Grande, Colorado, Snake, Columbia, Willamette, Sacramento, San Joaquin, St. Lawrence, Fraser, Okanagan

    b. *Water bodies*: Chesapeake Bay, Gulf of Mexico, San Francisco Bay, Puget Sound, Great Salt Lake, Lake Superior, Lake Michigan, Lake Huron, Lake Erie, Lake Ontario, Bay of Fundy, Gulf of St. Lawrence, Lake Winnipeg, Juan de Fuca Strait

    c. *Land bodies*: Cape Cod, Long Island, Delmarva Peninsula, Cape Hatteras, Florida Keys, Mississippi Delta, Mojave Desert, Olympic Peninsula, Vancouver Island, Grand Canyon, Alaskan Peninsula, Aleutian Islands, Newfoundland

    d. *Mountains*: Laurentians, Appalachians, Green Mountains, White Mountains, Adirondacks, Great Smoky Mountains, Ozark Plateau, Rocky Mountains, Wasatch Mountains, Sierra Nevada, Cascades, Klamath Mountains, Olympic Mountains, Pacific Coast Ranges, Black Hills

2. On the second map, political-cultural information should be entered as follows:

   a.   Write in the name of each U.S. State and Canadian province.

   b.   Draw in the major concentrations of minority populations, using Fig. 3-12 (p. 169) as a guide.

3. On the third outline map, urban-economic information should be entered as follows:

   a.   *Cities*: (locate and label with the symbol ●): Boston, Hartford, New York, Buffalo, Philadelphia, Pittsburgh, Baltimore, Washington, D.C., Richmond, Norfolk, Raleigh, Charlotte, Atlanta, Charleston (S.C.), Savannah, Jacksonville, Tampa, Orlando, Miami, Mobile, Birmingham, New Orleans, Memphis, Nashville, Louisville, Cincinnati, Columbus, Cleveland, Indianapolis, Detroit, Milwaukee, Chicago, Minneapolis-St. Paul, St. Louis, Des Moines, Kansas City, Omaha, Oklahoma City, Tulsa, Houston, Dallas-Ft. Worth, San Antonio, Austin, El Paso, Albuquerque, Denver, Salt Lake City, Tucson, Phoenix, Las Vegas, San Diego, Los Angeles, San Francisco, Portland (Ore.), Seattle, Vancouver, Calgary, Regina, Winnipeg, Windsor, Toronto, Ottawa, Montreal, Quebec City, Halifax

   b.   *Economic regions* (identify with circled letter):

   >   A -   Silicon Valley
   >   B -   Atlantic Seaboard Megalopolis
   >   C -   Corn Belt
   >   D -   Main Street conurbation
   >   E -   Pacific Hinge
   >   F -   Dairy Belt
   >   G -   Alaskan North Slope
   >   H -   Research Triangle
   >   I -   Boundary between Arid and Humid America (draw in)

PRACTICE EXAMINATION

Short-Answer Questions

Multiple-Choice

1. Which of the following is not a "corner" city of the American Manufacturing Belt (Continental Core Region)?

      a) St. Louis   b) Boston   c) Milwaukee
      d) Philadelphia   e) Baltimore

2. Which city is located closest to the Canadian capital of Ottawa?

      a) Toronto   b) Windsor   c) Vancouver
      d) Detroit   e) Washington, D.C.

3. Which of the following is a secondary economic activity?

      a) iron mining   b) beer brewing   c) retail sales
      d) managing a corporation e) cotton farming

4. In which region is Las Vegas located?

      a) Pacific Hinge   b) Western Frontier c) Southwest
      d) Continental Core   e) Northern Frontier

5. Which of the following is Canada's leading Pacific-coast city?

      a) Vancouver   b) Winnipeg   c) Alberta
      d) Windsor   e) Seattle

6. Which State lies entirely outside of the Continental Core Region?

      a) New York   b) New Jersey   c) Maine
      d) Michigan   e) Maryland

True-False

1. Manitoba is one of Canada's Prairie Provinces.

2. California's Silicon Valley is a good example of a *technopole*.

3. The shaping of the spatial form of the American metropolis during the 1920s was more heavily influenced by private automobiles than public transit systems.

4. Florida belongs to the South region of North America.

71

5. California lies entirely within Arid America.

6. Each of the Great Lakes is bordered by Canada.

Fill-Ins

1. The Atlantic Seaboard Megalopolis stretches northeastward from Washington, D.C. to the vicinity of the metropolitan area of _____.

2. The northward extension of the Sierra Nevada through the states of Oregon and Washington is called the _____ Mountains.

3. The outlet to the sea for the Great Lakes is the _____ River.

4. Alaska lies wholly within the North American region called the _____.

5. Denver and Salt Lake City lie on opposite sides of the physiographic province known as _____.

6. The newest subnational political entity within Canada, a huge territory created in 1999 that covers much of the country's northeast, is called _____.

Matching Question on States and Provinces

_____	1.	Atlantic province	A.	Saskatchewan
_____	2.	Corn Belt	B.	Texas
_____	3.	New England culture hearth	C.	California
_____	4.	Grand Canyon	D.	New York
_____	5.	North Slope oilfields	E.	British Columbia
_____	6.	High-tech complex triangle	F.	Virginia
_____	7.	Long Island	G.	Nova Scotia
_____	8.	San Andreas fault	H.	Quebec
_____	9.	Francophone Canada	I.	Illinois
_____	10.	Prairie province	J.	Ottawa
_____	11.	Canadian capital	K.	Massachusetts
_____	12.	Winter Wheat Belt	L.	Kansas
_____	13.	Vancouver	M.	Arizona
_____	14.	Capital Beltway	N.	Louisiana
_____	15.	Mississippi Delta	0.	Alaska

Essay Questions

1. The point was made that one of North America's greatest achievements was to overcome the "tyranny" of distance. Discuss how the United States and Canada overcame their physical barriers to allow their transcontinental economies and societies to emerge. Compare and contrast this experience in effective long-distance spatial organization with that of Russia since 1900 (referring back to the preceding chapter if necessary).

2. Draw a sketch map of the agricultural regions of the United States, and explain its spatial structuring within the framework of the macro-Thünian model.

3. Imagine that you are the chief executive of a successful young company that manufactures software for the newest high-speed computers. Where would you locate your company and why? Establish a "short list" of three possible sites, and choose one by a process of elimination in which your decision is built upon the most important locational variables for your plant and its highly-skilled work force.

4. Most of the western half of the United States is part of Arid America. Explain why this dry environment exists, and what its consequences are for climate, vegetation, and soils. Also, discuss the implications of this situation for the human use of the Earth, and speculate about what kind of economy might have arisen had the country been settled from the *Pacific coast* moving inland to the east.

5. The American central city has undergone a steady transformation over the past two centuries. Discuss its spatial evolution through the four stages of the Adams model. Then, briefly review the declining role of the big city in the age of the postindustrial multi-centered metropolis, stressing the context of the outer city, suburban downtowns, the *urban realms* model, and so-called residential reinvestment schemes in CBDs.

TERM PAPER POINTERS

The "Term Paper Pointers" section of the Introduction chapter in this **Study Guide** offered suggestions about approaching research and writing on geographic realms and their components, and you may wish to consult that material if you are undertaking a report on a North American region.

You should also be aware of the textbook's Web Site, which includes direct links to a number of Web Sites that may be quite helpful in your research. It is:
http://www.wiley.com/college/regions2000

Once again, the complexity of North America makes a good regional geography an appropriate place to begin your reading. Among the current titles recommended from the Chapter 3 citations in the book's **References and Further Readings** section are: Birdsall et al., Getis & Getis, McCann (Canada only), McKnight, Paterson, and Putnam & Putnam

(Canada only). Janelle offers a wide variety of regional vignettes. A goodly number of atlas-type publications are also worth consulting for certain topics, with many containing substantial texts in addition to map collections: Allen & Turner, *Atlas of North America* (National Geographic Society, 1985), Dean et al., *Historical Atlas of the United States* (National Geographic Society, 1988), Homberger, Riebsame, and Rooney et al. Several helpful topical treatments can be found as well. For physical geography see Atwood Espeland, and Hunt. Population geography (including migration) is treated by Boyle et al., Gober, Magocsi, Martin & Widgren, Pandit & Withers, Roberts, and Ward. Cultural and social geography are covered in Allen & Turner, Bell, Berry, Garreau (1981), Gastil, Kaplan, Krim, Lemco, Magocsi, Meinig (all 3 titles), Rooney et al., Shelley, and Zelinsky (both titles). Economic geography (including the postindustrial transformation) is treated in "A Survey of the Silicon Valley," Britton, Castells & Hall, Clark, Garreau (1991), Hart, Hartshorn (1997), Markoff, Pandit & Withers, "Telecom in North Texas," Wheeler et al., and Wilson. Historical geography is covered in Borchert (1967), Earle, Krim, Meinig (all 3 titles), Mitchell, Vance , Ward, and Zelinsky (both titles). Urban geography topics are treated in Adams, Berry, Borchert (both titles), Castells & Hall, Garreau (1991), Gottmann, Hartshorn (1992), Kaplan, Muller (both titles), Vance, Ward, and Yeates (both titles). Regional studies of subnational areas are treated in Bone, Doran, Gastil, Gottmann, Harris, Hartshorn (1997), Hunt, Krim, Lemco, Magocsi, Rimer, Webster, and Yeates (1975).

NORTH AMERICA

| 0 | 500 | 1000 | 1500 Kilometers |

| 0 | 300 | 600 | 900 Miles |

Arctic Circle

Tropic of Cancer

NORTH AMERICA

| 0 | 500 | 1000 | 1500 Kilometers |
| 0 | 300 | 600 | 900 Miles |

180°

70°

80°

70°

20°

160°

40°

50°

50°

40°

140°

60°

50°

30°

30°

30°

120°

100°

Arctic Circle

Tropic of Cancer

80°

NORTH AMERICA

	500	1000	1500 Kilometers
0	300	600	900 Miles

Arctic Circle

Tropic of Cancer

180°
70°
80°
70°
160°
20°
40°
50°
50°
140°
60°
30°
30°
120°
100°
80°

NORTH AMERICA

500 1000 1500 Kilometers
300 600 900 Miles

Arctic Circle

Tropic of Cancer

CHAPTER 4

MIDDLE AMERICA

OBJECTIVES OF THIS CHAPTER

Chapter 4 is a survey of Middle America (Mexico, Central America, and the Greater and Lesser Antilles that constitute the islands of the Caribbean Basin). Following an introduction, the sequential influences of the Amerindian (Mesoamerican) civilizations and the Hispanic colonizers are reviewed, and their various cultural collisions are evaluated in the shaping of contemporary society. The regional structure of the realm is initially presented within the useful "Mainland-Rimland" framework (set against the complex European colonial legacy). Each major region is then treated: the Caribbean—underscoring cultural and economic geography and the impact of tourism; Mexico—highlighting social and economic geography as well as plans for continuing development; and Central America—profiling each of its 7 republics, and emphasizing spatial dimensions of the instabilities that continue to plague this region.

Having learned the regional geography of Middle America, you should be able to:

1. Understand which components make up the Middle American realm and their differing physiographies.

2. Describe the major contributions of the Maya, Aztecs, and Spaniards in shaping the contemporary cultural and social geography of the realm (and the African and other European colonial infusions in much of the Rimland portion of Middle America).

3. Differentiate between Mainland and Rimland Middle America in terms of political, cultural, and economic regional geography.

4. Understand the geographic patterns of the Caribbean Basin, its evolution, and its prospects for future change.

5. Understand the geography of Mexico, especially its development opportunities as well as its problems related to economic inequalities and population growth.

6. Understand the altitudinal zonation of environments that mark the economic and settlement geographies of Middle and South America.

7. Understand the geographic patterns of Central America, including the challenges facing each of its republics.

8. Locate the major physical, cultural, and economic-spatial features of the realm on an outline map.

GLOSSARY

Isthmus (202)

Performs the same function as a land bridge, but is usually shorter in length and narrower in width. On the mainland Middle American land bridge, the South American end is the isthmus of Panama.

Land Bridge (202)

A narrow isthmian link that connects two larger landmasses, such as the 3800-mile-long (6000-km-long) mainland of Middle America that connects North and South America between the U.S. border and the southeastern end of the Panamanian isthmus.

Greater Antilles (203)

The larger islands of the northern Caribbean that encompass Cuba, Hispaniola (containing Haiti and the Dominican Republic), Puerto Rico, and Jamaica.

Lesser Antilles (203)

The smaller-island arc of the eastern Caribbean, stretching southward from the Virgin Islands to Trinidad near the South American coast. Region can also be extended north-westward toward Florida to include the Bahamas island chain.

Archipelago (203)

A set of islands grouped closely together, usually elongated into a chain.

Culture hearth (203)

A source area or innovation center from which cultural traditions are transmitted.

Mesoamerica (203)

Anthropological label for the Middle American culture hearth, shown in world context in Fig. 6-3.

Plaza (206)

Central market square that was the focus of Spanish New World towns, containing church and government buildings.

Mainland-Rimland framework (208-209)

Augelli's framework that recognizes a Euro-Amerindian Mainland and a Euro-African Rimland in Middle America, as mapped in Fig. 4-6 on p. 208.

Hacienda (209)

Literally, a large estate in a Spanish-speaking country. Sometimes equated with plantation, but there are important differences between these two types of agricultural enterprise and rural land tenure. Many have now been divided into smaller holdings or reorganized as cooperatives.

Plantation (209-210)

A large estate owned by an individual, family, or corporation and organized to produce a cash crop. Almost all plantations were established within the tropics.

Ejido (210)

Communally-owned cooperative farmlands in central and southern Mexico; former *hacienda* lands.

Mulatto (216)

A person of mixed African (black) and European (white) ancestry.

Mestizo (216)

A person of mixed white and Amerindian ancestry.

Plural(istic) society (216)

A society in which two or more population groups, each practicing its own culture, live adjacent to one another without mixing inside a single state.

Acculturation (220)

Cultural modification resulting from intercultural borrowing. In cultural geography, the term refers to the change that occurrs in the culture of indigenous peoples when contact is made with a society that is technologically superior.

Transculturation (220)

Two-way cultural borrowing that occurs when different cultures of approximately equal complexity and technological level come into close contact. In *acculturation*, by contrast, an indigenous society's culture is modified by contact with a technologically more advanced society.

Maquiladora (224-225)

The term given to modern industrial plants in Mexico's northern (U.S.) border zone. These foreign-owned factories assemble imported components and/or raw materials, and then export finished manufactures, mainly to the United States. Most import duties are minimized (and will be phased out under NAFTA by 2002), bringing jobs to Mexico and the advantages of low wage rates to the foreign entrepreneurs.

"Dry Canal" (225)

An overland rail and/or road corridor across an isthmus dedicated to performing the transit functions of a canalized waterway. Best adapted to the movement of containerized cargo, there must be a port at each end to handle the necessary *break-of-bulk* unloading and reloading.

Altitudinal zonation (228-box)

Vertical regions defined by physical-environmental zones at various elevations, particularly in the highlands of South and Middle America. See Fig 4-13.

Tierra caliente (228-box)

The lowest of five vertical zones into which the settlement of highland Middle and South America is divided according to elevation. The *caliente* is the hot humid coastal plain and adjacent slopes up to 2500 feet (750 meters) above sea level. The natural vegetation is the dense and luxuriant tropical rainforest; the crops are tropical, including bananas. See Fig. 4-13.

Tierra templada (228-box)

The second altitudinal zone in highland Middle and South America, between 2500 and 6000 feet (750 and 1850 meters). This is the "temperate" zone, with moderate temperatures compared to the *tierra caliente*. Crops include tobacco, coffee, corn, and some wheat. See Fig. 4-13.

Tierra fría (228-box)

The third altitudinal zone in highland Middle and South America, from about 6000 feet (1850 meters) up to the tree line at nearly 12,000 feet (3600 meters). Coniferous trees stand here; upward they change into scrub and grassland. There are also important pastures within the *fría*, and wheat, potatoes, and barley can be cultivated. See Fig. 4-13.

Tierra helada (228-box)

The fourth settlement zone in highland South America, extending upward from about 12,000 to 15,000 feet (3600-4500 meters). This altitudinal zone lies above the tree line, and is so cold and barren that it can only support the grazing of hardy livestock and sheep. See Fig.4-13.

Tierra nevada (228-box)

This zone is above the snow line, which lies at approximately 15,000 feet (4500 meters), and is referred to as the "frozen land." See Fig. 4-13.

Tropical deforestation (234-box)

The clearing and destruction of tropical rainforests to make way for expanding settlement frontiers and the exploitation of new economic opportunities.

SELF-TESTING QUESTIONS

Cover the right side of the page with a sheet of paper. Uncover each line after you have attempted to answer the question in the left column. If necessary, refer to textbook page(s) listed at the right.

Question	Answer	Page

Middle American Characteristics

What is the spatial extent of this realm?	Mexico and Central America—from the U.S. border to the northern edge of South America— plus all of the Caribbean islands to the east.	201
Why is Middle America more culturally diverse than South America?	African and Asian ancestries prevail beside those of European background; the Amerindian cultural contribution is greater; the Caribbean is a region of especially complex cultural pluralism.	202
What is a *land bridge*?	A narrow isthmian link between two large landmasses.	202
What is the difference between Central and Middle America?	*Central America* is the mainland between Mexico and South America, containing the 7 republics of Belize, Guatemala, Honduras, El Salvador, Nicaragua, Costa Rica, and Panama. Besides all of Central America, *Middle America* also includes Mexico and the Caribbean Basin.	203-box

Mesoamerican Legacy

What is *Mesoamerica*?	The Middle American culture hearth that stretched from northern Mexico southeast to central Nicaragua.	203
Where was the Maya civilization centered?	In Guatemala and the Yucatán Peninsula of Mexico.	204
Describe some of the Maya's accomplishments.	Urbanization, pyramids, spectacular palaces, stone carvings and other artwork, mathematics, astronomy, calendrics, advanced agriculture, and a wide trading network.	204
Who were the Aztecs?	The successors to the Toltecs, who succeeded the Maya.	204
What was the central area of the Aztec's empire?	The Aztec state centered in the Valley of Mexico, headquartered at the city of Tenochtitlán.	204

79

Which crops did the Amer-indians of Meso-america contribute to the world?	Corn (maize), various kinds of beans, the sweet potato, the tomato, cacao, squash, and tobacco.	205
What kinds of agriculture were emphasized after the Spanish conquest?	The keeping of livestock, especially cattle and sheep—which competed with the growing of the subsistence crops of the conquered Amerindians.	205
What was the most far-reaching change in the cultural landscape brought by the Spaniards?	The resettlement of the Amerindians from rural land into villages and towns laid out and controlled by the conquerors; these towns were administrative centers for tax collection and labor recruitment, especially for mining.	206
How were Spanish New World towns organized?	The focus of the town was the central plaza, where church and government buildings were located. The surrounding streets were arranged in an easily defensible gridiron pattern.	206
What further changes were engendered by the mining industry?	The creation of a network of tightly-controlled Spanish towns that cemented power over even the more isolated parts of Mexico.	206-207

Mainland and Rimland

What is the difference between the cultures of Mainland and Rimland Middle America?	The Mainland is dominated by a Euro-Amerindian cultural heritage; the Rimland is dominated by a Euro-African cultural heritage.	208-209
What are the geographic differences between Mainland and Rimland?	The Rimland was an area of sugar and banana plantations, high accessibility, seaward exposure, and maximum cultural contact and mixture; the Mainland was removed from these contacts, the region of the hacienda, more self-sufficient, and less dependent on outside markets.	209
What are the five characteristics of Middle American plantations?	They are located in the tropical coastlands and islands; they produce single export crops; foreign ownership and profit outflow dominate; labor is seasonal and has often been imported; "factory in the field" methods are far more efficient than those of the hacienda.	210

What are the political legacies of Middle American colonialism?	The Mainland states are now independent republics, but all except Belize have Hispanic origins; the cultural variety of the Caribbean is much greater— Cuba, Puerto Rico, and the Dominican Republic were Spanish; Haiti and several islands of the Lesser Antilles were French; Jamaica, Trinidad, and many lesser islands were British; and other small islands had Dutch, Danish, and even U.S. affiliations.	210-211

The Caribbean Basin

Which two ethnic groups suffered as a result of the European sugar trade?	The Amerindians were virtually wiped off the map, and Africans were imported in bondage.	211
Why do the export crops and the mineral resources of the islands provide so modest an income for the region, and why does wide-spread poverty persist?	These commodities face severe competition from many other disadvantaged countries, and are not established on a scale that could improve local living standards; most islanders therefore live in poverty and eke out a subsistence existence from a small plot of poor land.	212-215
What is a *mulatto*?	A person of mixed white-African (black) ancestry.	216
What is a *mestizo*?	A person of mixed white-Amerindian ancestry.	216
Why is there a large population of Asian peoples in the Caribbean?	During the 19th century, emancipation of slaves and ensuing labor shortages brought immigrants (mostly indentured servants) from distant locations, including China and India.	216
Which islands have size-able Asian populations?	Trinidad, Jamaica, Guadeloupe, and Martinique.	216
Why is tourism a mixed blessing for the Caribbean?	It provides revenues and jobs in a region of limited options; however, the fabric of local communities are strained by the disparity and contrast of opulence and poverty in the cultural landscape.	216-217

Mexico

How strong is the Amerindian imprint on Mexican culture?	Extremely strong: 60 percent of the population are mestizos, 30 percent are Amerindian, and only 9 percent are European.	220

Did the 1910 Revolution achieve its goals?	For the most part, yes: the haciendas were redistributed, many into communally-owned *ejidos*; the Revolution also resurrected the Amerindian cultural contribution to Mexican life.	220-221
Where is most of Mexico's oil production located?	Along the central and southern Gulf Coast, especially around the city of Villahermosa and the nearby Bay of Campeche.	224
What are *maquiladoras*?	Modern industrial plants in Mexico's northern (U.S.) border zone. These foreign-owned factories assemble imported components and/or raw materials, and then export finished manufactures, mainly to the United States. Most import duties are minimized, bringing jobs to Mexico and the advantages of low wage rates to the foreign entrepreneurs.	224-225

Central America

Name the seven republics of Central America.	Guatemala, Honduras, Belize, El Salvador, Nicaragua, Costa Rica, and Panama. See Fig. 4-12.	226-227; 227-map
Environmentally, where is most of the region's population concentrated?	In the *templada* zone of the highlands, toward the Pacific side of the land bridge.	226
Name the five altitudinal zones of human settlement in highland Middle and South America.	*Tierra caliente* (sea level to 2500 feet), *tierra templada* (2500-6000 feet), *tierra fria* (6000-12,000 feet), *tierra helada* (12,000-15,000 feet), and the *tierra nevada* (15,000 feet and above).	228-box
What basic conflicts divide the population of Central America?	Amerindian and mestizo population clusters often clash; also, there is a huge gulf between the privileged and the poor, causing resentment and violence.	228
How have Central America's prospects brightened since the early 1990s?	There is a new spirit of regional identity and cooperation, resulting in pacts for an economic union but hurricane Mitch in 1998 set back Honduras and Nicaragua for years.	229
Which republic was run by the Sandinistas during the 1980s? Which republic is the "Switzerland" of Central America?	Nicaragua; Costa Rica.	232; 232-233
What is the current status of the Panama Canal?	The U.S. withdrawal was completed at the end of 1999 and the Canal turned over to Panama.	235

MAP EXERCISES

Map Comparison

1. The map of Middle America's regions (Fig. 4-2, pp. 200-201) clearly offsets each of the geographic components of this realm. Compare and contrast the environments of these regions, basing your observations on the world maps of landscapes (Fig. I-1), precipitation (Fig. I-7), and climates (I-8) in the Introduction chapter.

2. Compare Figs. 4-5 and 4-6 (pp. 207, 208) and make observations about the colonial influence on the Mainland/Rimland split. Why is the Central American east coast included in the Rimland rather than the Mainland region? What generalizations can be made about the composition of present Caribbean populations based on the European countries that once ruled them?

3. The most significant recent development in Mexico's manufacturing geography is the rise of the *maquiladoras*. Compare the maps of maquiladora location (Fig. 4-11), raw materials (Fig. 4-9), and the internal States of Mexico (Fig. 4-10). Comment on the advantages that the northern border holds for industrial activity, and on what the highly clustered location of maquiladoras means for the various regional economies of Mexico.

Map Construction   *(Use outline maps at the end of this chapter)*

1. In order to familiarize yourself with Middle American physical geography, place the following on the first outline map:

   a. *Water bodies*: Caribbean Sea, Gulf of Mexico, Bay of Campeche, Gulf of California (Sea of Cortés), Gulf of Tehuantepec, Gulf of Honduras, Gulf of Panama, Gulf of Darien, Gulf of Fonseca, Panama Canal, Rio Grande River, Lake Nicaragua, Straits of Florida, Windward Passage

   b. *Land bodies and features*: Greater Antilles, Lesser Antilles (including the Bahamas Islands), Leeward Islands, Windward Islands, Virgin Islands, Baja California, Yucatán Peninsula, Isthmus of Panama, Hispaniola, Florida Keys, Sierra Madre Occidental, Sierra Madre Oriental, Valley of Mexico, Sonora Desert, Isthmus of Tehuantepec

2. On the second map, enter the name of every country that is shown, including all of the mainland republics and as many island-states as possible.

3. On the third outline map, urban-economic information should be entered as follows:

   a. *Capital cities* (locate and label with the symbol *): Mexico City, Belmopan, Guatemala City, San Salvador, Tegucigalpa, Managua, San José, Panama City, Havana, Kingston, Nassau, Port-au-Prince, Santo Domingo, San Juan, Fort-de-France, Port of Spain, Willemstad

b.   *Other cities* (locate and label with the symbol ●): Ciudad Juarez, Tijuana, Monterrey, Torreón, Chihuahua, Durango, Zacatecas, Guadalajara, Tampico, Veracruz, Acapulco, Oaxaca, Villahermosa, Mérida, Cozumel, Belize City, Quezaltenango, San Pedro Sula, San Miguel, León, Bluefields, Limon, Colón, Santiago de Cuba, Mariel, Guantanamo, Montego Bay, Puerto Plata, Mayaguez

c.   *Economic Regions* (identify with circled letter):

A - Curaçao
B - Major area of *ejidos* today
C - Bay of Campeche oilfield
D - Panama Canal
E - Mexican gold placering area
F - Maquiladora zone

PRACTICE EXAMINATION

Short Answer Questions

Multiple-Choice

1. The racial term applied to people of mixed white and Amerindian ancestry is:

    a) *mulatto*   b) *caliente*   c) *ejido*
    d) *mestizo*   e) *amerindiano*

2. In which mainland republic did the Sandinistas overthrow the Somoza regime in the late 1970s?

    a) Nicaragua   b) El Salvador   c) Honduras
    d) Mexico   e) Dominican Republic

3. The most populous of all the Middle American countries is:

    a) Nicaragua   b) Panama   c) Mexico
    d) Honduras   e) Guatemala

3. Which Mainland country does not contain a portion of the Rimland?

    a) El Salvador   b) Belize   c) Honduras
    d) Nicaragua   e) Panama

4. Which part of Mexico is most closely associated with the ancient Maya culture?

    a) Baja California   b) Valley of Mexico   c) the Yucatán Peninsula
    d) Hispaniola   e) Rio Grande Valley

5. Which Central American capital is located on the Pacific?

a) San José   b) Mexico City   c) Belmopan
d) Guatemala City   e) Panama City

True-False

1. The core area of the Aztec state was located in what is still the core area of Mexico today.

2. The large island of Trinidad is located in the Greater Antilles.

3. Costa Rica is a U.S. enemy that is a haven for communist insurgencies throughout Middle and South America.

4. The Chiapas rebellion occurred in southern Mexico.

5. The *tierra nevada* altitudinal zone does not occur in Central America.

6. Before Castro's revolution, Cuba was a colony of the United States.

Fill-Ins

1. The gridiron street pattern was first introduced to Middle America by Europeans from the country of _____.

2. The lowest-lying altitudinal zone of agricultural activity, extending from sea level to an elevation of 2500 feet (750 meters), is the *tierra* _____.

3. The Yucatán Peninsula is a part of the country of _____.

4. The largest island of the Greater Antilles is _____.

5. The country most widely devastated by Hurricane Mitch in 1998 was _____.

6. The only Middle American country with a direct land link to the South American continent is _____.

Matching Question on Middle American Countries

_____ 1. U.S. backed government in recent civil war

_____ 2. Former bastion of the Sandinistas

_____ 3. "Switzerland of Central America"

_____ 4. Large East Indian population

_____ 5. Eastern half of Hispaniola

_____ 6. Former Danish colony

_____ 7. Possibility of U.S. statehood

_____ 8. The largest Caribbean island

_____ 9. Bauxite (aluminum ore) mining

_____ 10. Formerly belonged to Colombia

_____ 11. Still a Dutch colony

_____ 12. Western half of Hispaniola

_____ 13. Hit the hardest by Hurricane Mitch

_____ 14. Part of Maya culture hearth

_____ 15. Formerly British Honduras

A. Mexico
B. Puerto Rico
C. Dominican Republic
D. Curaçao
E. Jamaica
F. U.S. Virgin Islands
G. Honduras
H. Haiti
I. Belize
J. Panama
K. Nicaragua
L. Costa Rica
M. Cuba
N. Trinidad and Tobago
O. El Salvador

Essay Questions

1. One of this realm's most important regional frameworks is the Mainland/Rimland scheme. Compare and contrast the physical, economic, and cultural geographies of each of these components of Middle America, and show how its historical evolution proceeded in a direction that poses a challenge to regional integration and unity in the future.

2. Discuss the European impact on the shaping of the cultural and political geographies of the Caribbean Basin today. Why is this region trapped in a cycle of poverty that is rooted in the economic system inherited from the colonial era?

3. Describe the fivefold vertical zonation of settlement environments in highland Middle and South America, including the agricultural geography of each zone.

4. Few countries in the developing world possess the opportunities and challenges facing Mexico. Discuss those aspects of Mexican geography that offer a real potential for meaningful progress in living standards in the foreseeable future, and weigh them against the problems the country confronts in the short- and longer-term future.

5. Much of Central America's turmoil has been rooted in this region's past. Trace the historical geography of the region, paying attention to Amerindian, Spanish, and other cultural influences. What new economic prospects brighten Central America's future, and what obstacles must be overcome in order for progress to occur?

TERM PAPER POINTERS

The "Term Paper Pointers" section of the Introduction chapter in this **Study Guide** offered suggestions about approaching research and writing on geographic realms and their components, and you may wish to consult this material if you are undertaking a report on a Middle American region.

You should also be aware of the textbook's Web Site, which includes direct links to a number of other Web Sites that may be quite helpful in your research. It is: http://www.wiley.com/college/regions2000

The only full-length geography of this entire realm is cited in the book's **References and Further Readings** section: West, Augelli et al., a somewhat dated but still comprehensive survey that is practically indispensable for doing a paper on Middle America. Five additional, but more general, overviews of both Middle and South America may prove helpful too— Blakemore & Smith, Blouet & Blouet, Clawson, Gilbert (1990), and James & Minkel. For regions within the realm, see Barker & McGregor, Boswell & Conway, Elbow, Herzog, Klak, Knight, Lowenthal, MacPherson, Paige, Pattullo, Pick & Butler (1993), Portes & Grosfoguel, Richardson, and Sealey. The political turmoil of this realm is covered in Barton, "Cuba...," Grossman, Harvey, Klak, Knight, J. Preston, Rohter (Nov. 1998), and Trias Monge. Myers offers a thorough background on the tropical rainforest environment. The historical geography of the realm is treated in Augelli, Collier, Davidson & Parsons, Greenfield, McCullough, Paige, Richardson, Sargent, Watts, and West. On special topics, regionalization is treated by Augelli; the development process in Barker & McGregor, Elbow, Greenfield, Kopinak, Momsen, D. Preston, Richardson, Watts, and West; the U.S.-Mexico border zone in Arreola & Curtis, Herzog, Kopinak, MacLachlan & Aguilar, "Maquiladora Industry," and "Mexico's New Frontier." The increasingly important topic of urbanization is covered in Arreola & Curtis, Gilbert (1996), Greenfield, Griffin & Ford, Herzog, Momsen, Pick & Butler (1997), Potter, Sargent, and Ward.

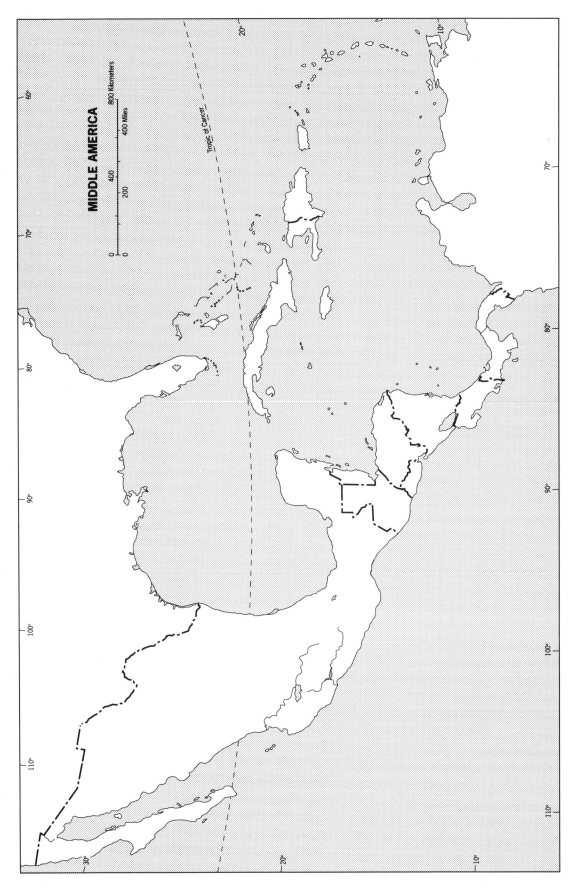

MIDDLE AMERICA

Tropic of Cancer

0   200   400   400 Miles
0   400   800 Kilometers

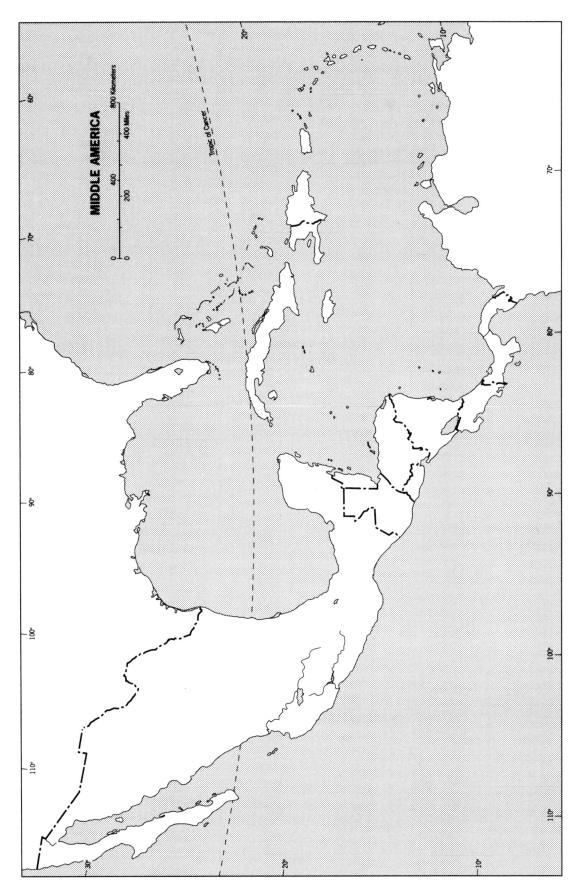

MIDDLE AMERICA

Tropic of Cancer

400
200
400 Miles

800 Kilometers

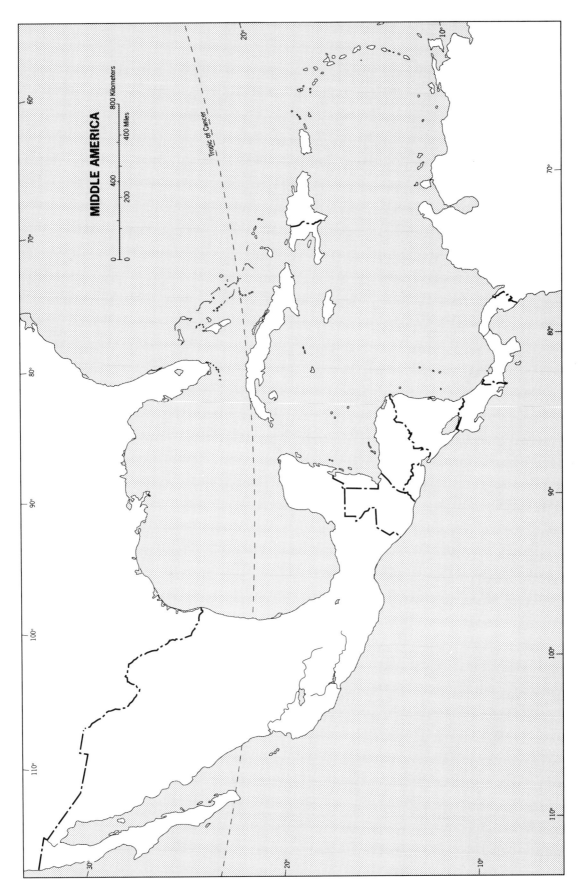

MIDDLE AMERICA

200 400

0 400 800 Kilometers

0 400 Miles

Tropic of Cancer

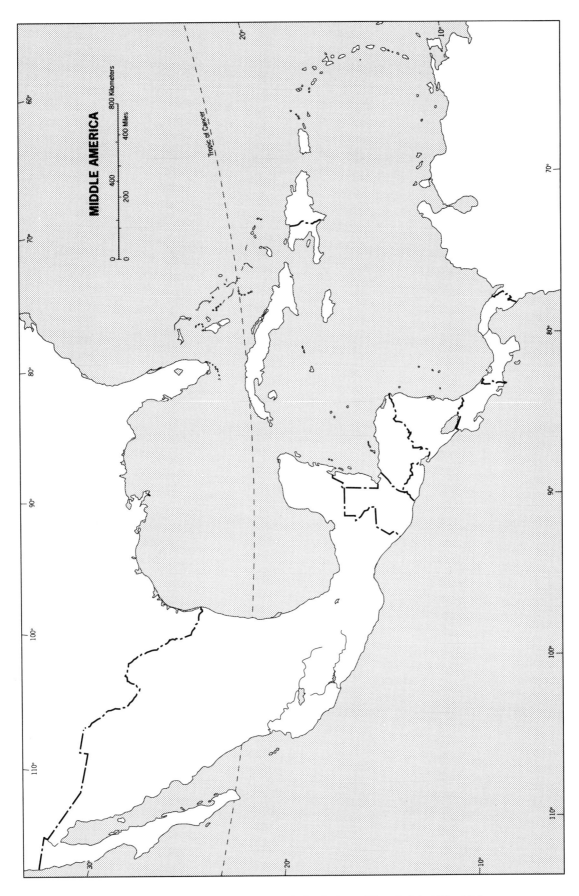

MIDDLE AMERICA

800 Kilometers

400 Miles

400
200

0
0

Tropic of Cancer

CHAPTER 5
SOUTH AMERICA

OBJECTIVES OF THIS CHAPTER

Chapter 5 treats South America, a realm of vast potential but also of frustrations in the effort to move itself forward. After a general introduction to the chapter, South America's historical geography is traced in some detail, with emphasis on its component culture areas. Urbanization is introduced, highlighting a model of "Latin" American city structure. A review of South America's regions completes the chapter, presented within a four-part framework covering giant Brazil, the Caribbean North, the Andean West, and the mid-latitude Southern Cone.

Having learned the regional geography of South America, you should be able to:

1. Understand the major physiographic features of the realm.

2. Grasp the essentials of the historical geography of this continent.

3. Understand the various influences that have shaped South America's culture areas.

4. Deepen your understanding of Middle and South American urbanization, building on the Griffin-Ford model introduced on text pp. 251-253.

5. Describe the broad regionalization pattern of the realm, with reference to cultural patterns, historical, and political developments.

6. Understand the geographic essentials of each South American republic, including its resources, agricultural patterns, natural environments, and economic development potential.

7. Locate the leading physical, cultural, and economic-spatial features of the realm on an outline map.

GLOSSARY

Isohyets (239)

Lines connecting places of equal precipitation totals.

Altiplano (241)

High-elevation plateau or basin between even higher mountain ranges; Andean altiplanos often lie at altitudes in excess of 10,000 feet (3000 m).

Land alienation (242)

The process by which Spanish invaders took over Amerindian lands in order to form large haciendas.

Plural society (245)

A society in which two or more populations groups, each practicing its own culture, live adjacent to one another without mixing inside a single state.

Commercial agriculture (246)

For-profit agriculture.

Subsistence agriculture (246)

Minimum-life-sustaining agriculture.

"Latin" American City model (251-253)

The Griffin-Ford model of Middle and South American intraurban spatial structure, discussed on pp. 251-253, and diagramed on p. 252.

Plaza (252)

The old hub and focus of the Middle and South American city, the open central square flanked by the main church and government buildings (see photo p. 252).

Informal sector (253)

Dominated by unlicenced sellers of goods and services, the primitive form of capitalism found in many developing countries that takes place beyond the control of government.

Barrio (favela) (253)

Term meaning "neighborhood" in Spanish. Usually refers to an urban community in a Middle or South American city; often slums are known as *barrios* or in Brazil as *favelas*.

Sertão (256)

The dry inland back-country of northeastern Brazil.

El Nino (257)

A periodic, large-scale, abnormal warming of the sea surface in the low latitudes of the eastern Pacific Ocean that has global implications, disturbing normal weather patterns in many parts of the world, especially South America.

Fazendas (258)

Large-scale coffee plantations.

Forward capital (260)

A capital city located near a sensitive zone with a neighboring state or a frontier that a country wishes to develop; such a statement concerning the push to the empty interior of Brazil was made in the 1950s, when the government decided to relocate its headquarters from Rio de Janeiro to Brasília.

Selvas (261)

Another term for tropical rainforests.

Growth pole (261)

An urban center with certain attributes that, if augmented by investment support, will stimulate regionwide economic development of its hinterland.

Llanos (263)

Savanna-like grasslands, especially of the low-lying Orinoco Basin in Venezuela as well as neighboring Colombia.

Tierra caliente *(see Chapter 4, p. 228-box)*

The lowest of four vertical zones into which the settlement of highland South America is divided according to elevation. The *caliente* is the hot humid lowland and adjacent slopes up to 2500 feet (750 m) above sea level. The natural vegetation is the dense and luxuriant tropical rainforest; the crops are bananas, sugar, cacao, and rice in the lower areas and coffee, tobacco, and corn along the somewhat higher slopes.

Tierra templada *(see Chapter 4, p. 228-box)*

The second altitudinal zone in highland South America, between 2500 and 6000 feet (750 and 1850 m). This is the "temperate" zone, with moderate temperatures compared to the *tierra caliente.* Crops include tobacco, coffee, corn, and some wheat.

Tierra fría *(see Chapter 4, p. 228-box)*

The third altitudinal zone in highland South America, from about 6000 feet (1850 m) up to nearly 12,000 feet (3600 m). Coniferous trees stand here; upward they change into scrub and grassland. There are also important pastures within the *fría*, and wheat can be cultivated. Several major population clusters in the Andes lie at these elevations.

Tierra Helada *(see Chapter 4, p. 228-Box)*

The fourth settlement zone in highland South America, extending upward from about 12,000 to 15,000 feet (3600-4500 m). This altitudinal zone is so cold and barren that it can only support the grazing of hardy livestock and sheep.

Tierra Nevada *(see Chapter 4, p. 228-Box)*

Above the snow line, lying at approximately 15,000 feet (4500 m), this is the uninhabitable "frozen land."

Insurgent state *(267)*

Territorial embodiment of a successful guerilla movement. The establishment by anti-government insurgents of a territorial base in which they exercise full control; thus a state within a state.

Oriente *(272-273)*

Literally "the east," refers to the jungly lowlands of Peru and Ecuador to the east of the Andes that are sparsely populated but contain petroleum deposits.

Pampa *(273-274)*

Literally the word means "plain." The physiographic subregion of east-central Argentina that is the country's leading crop-and-livestock-producing region.

Von Thunen model *(275)*

Explains the location of agricultural activities in a commercial economy. A process of spatial competition allocates various farming activities into concentric rings around a central market city, with profit-earning capability the determining force in how far a crop locates from the market. The original (1826) Isolated State model now applies to the continental scale (see Fig. 1-6).

Elongated State (277-278)

A state whose territory is decidedly long (at least six times longer than its width) and often creates external political, internal administrative, and general economic problems. Chile is a classic example.

SELF-TESTING QUESTIONS

Cover the right side of the page with a sheet of paper. Uncover each line after you have attempted to answer the question in the left column. If necessary, refer to textbook page(s) listed at right.

Question	**Answer**	**Page**
South American Characteristics		
How does South America's longitudinal position differ from North America's?	It lies considerably farther to the east, closer to Africa but facing a much wider Pacific Ocean on the west.	239
What is the realm's dominant physiographic feature?	The Andes Mountains, which form an imposing north-south barrier along South America's entire west coast.	239
Where are modern South America's largest population concentrations situated?	In the east and the north of the realm.	240
Historical Geography		
Who were the Incas?	Descendants of ancient peoples who created a major civilization in the northern Andes around A.D. 1200, centered in Peru's Cuzco Basin.	240-242
List some of the Incas' major achievements.	Most of all, the political integration of Andean South America; a splendid circulation system for goods and ideas; administration of a complex social and economic system; Quechua—a regional language that still survives.	241-242

How and when did the Spaniards conquer the Incas?	In 1533, Pizarro led a small band of soldiers into Cuzco and deposed the Inca Empire, which was ripe for overthrow given its internal divisions and problems of royal succession.	242
How did Spain and Portugal resolve their 15th century disputes over South American territory?	The 1494 Treaty of Tordesillas, mediated by the Pope, gave Spain all land west of the 50th meridian and Portugal all territory to the east.	243-244
Why did the Portuguese import so many Africans to coastal Brazil?	They opted for a Caribbean-style plantation economy, which required the services of millions of African slaves.	244
How did South America's 19th century independence movement spread?	Argentina and Chile, farthest from Peru, began it and Bolívar pushed down from New Granada to the north—by 1824, the Spaniards were driven out; in Brazil, the Prince Regent's son defied his father, proclaimed independence, and expelled the loyalist Portuguese forces in 1822.	245

Culture Areas

What makes the pattern of South American agriculture unusual?	Commercial and subsistence agriculture exist side by side to a greater degree than in any other realm (see Fig. 5-4).	246-box
Name the 5 internal culture spheres of this realm as defined by Augelli.	*Tropical plantation* region of the northeast and northern coasts; *European-commercial* region of most of the mid-latitude south; the *Amerind-subsistence* region of the Andean valleys and plateaus; the *Mestizo-transitional* region covering large parts of the interior between Amerindian and commercial agriculture; and the *Undifferentiated* primitive areas of the most isolated zone (largely Amazonia).	246-248
How well unified is the South American realm?	In the past, surprisingly little interaction occurred among the realm's countries, but there are signs of change as a new era of cooperation appears to be opening. Mutually advantageous trade is the catalyst for international cooperation.	248-249

Urbanization

How fast is South American urbanization increasing?	Very rapidly: in 1925, the realm was 33% urban, by 1950 40%, but over 60% in 1975 and about 76% today; the urban areas now increase annually by 5%, the rural areas by less than 2%.	249
Why does the migration toward cities remain so high in the 1990s?	*Push* factors are rural poverty and lack of land reform; reinforcing *pull* factors are perceived employment opportunities, education, medical care, and a more exciting pace of life.	251
Name the land-use zones of the "Latin" American city model.	The CBD, the commercial spine, the elite residential sector, the zone of maturity, the zone of *in situ* accretion, the zone of peripheral squatter settlements, and the disamenity sector containing the slums.	251-253

Brazil

How large is Brazil's relative size?	Territorially, it occupies just under 50% of South America; it ranks fifth in the world after Russia, Canada, China, and the United States.	254
What centripetal forces bind Brazil together?	Successful ethnic mixing, adherence to the Catholic faith, a common language, and a cultural heritage of common music, fashion, and art.	254
How has the Brazilian economy fared since 1980?	Ups and downs; the great promise of a "takeoff" has been slowed, in recent years, due to regional economic and social inequities.	256
What is Brazil's politico-geographical framework?	A federal republic consisting of 26 States, and the federal district of Brasília.	256
What is the growth pole concept?	A development plan in which a set of "seed" industries are nurtured, thereafter setting off "ripples" of growth in the surrounding area.	261

Caribbean South America

What forces bind this region?	Common coastal location; the legacy of a tropical plantation culture and economy; large black and Asian minorities.	262
What is the political status of each of the Guianas?	Guyana is independent, with unstable leadership; Suriname, also unstable, is now independent of the Dutch; French Guiana remains an overseas *département* of France.	264-box

What are Venezuela's major activities?	Lake Maracaibo oil, and iron ores in the east; oil reserves have also been discovered in the *llanos* regions of both Venezuela and Colombia.	263-264
Where is most of Colombia's coffee grown?	In the *tierra templada* zone of the Andean slopes, centered in the provinces of Antioquia and Caldas. The industry is now threatened, however, by many serious problems, and its survival is in doubt.	266-267
What is an insurgent state?	The territorial base in which a guerilla movement exercises full control; therefore, a state within a state.	267

The Andean West

Which states make up this region? Which culture dominates? Which groups rule?	Peru, Ecuador, and Bolivia; the Amerind-subsistence cultural sphere is dominant, with a large mestizo population besides the majority Amerindians; none-theless, the European elite holds most of the political power.	268
What are the three physiographic/cultural subregions into which Peru is divided?	(1) The desert coast, the European-mestizo region; (2) the Andean Highlands or Sierra, the Amerindian region; (3) the eastern slopes and montaña, the sparsely populated Amerindian-mestizo interior.	269
Name the five altitudinal zones of human settlement in Andean South America.	*Tierra caliente* (sea level to 2500 feet), *tierra templada* (2500-6000 feet), *tierra fría* (6000-12,000 feet), and the *puna (páramos)* (12,000-15,000 feet). For a fuller discussion of the altitudinal zonation scheme, see box on text p. 228.	271; see also ch.4, p. 228-box
What is the Sendero Luminoso?	The Peruvian radical-communist insurgency that threatened the stability of the country for nearly two decades. The guerrillas have been in retreat since the mid-1990s, following a strong governmental counterattack.	271
What crucial territory did Bolivia lose to Chile?	Bolivia lost its Pacific outlet in the late 19th century, which leaves the country in a landlocked, disadvantageous situation.	273

The Southern Cone

What is the significance of the Argentine Pampa?	Productive meat-and-grain region which has made the country a major food exporter; gave Buenos Aires a rich hinterland, spurring its growth and success.	273-275

What is the status of the Falkland Islands in the 1990s?	The British rule following their 1982 ouster of the Argentinean military; further negotiations are needed, because unresolved sovereignty claims could trigger another conflict.	276
Compare and contrast the population distributions of Argentina, Uruguay, and Chile.	Argentina has a strongly peripheral pattern; Uruguay's population is uniformly spread across its territory; elongated Chile has a highly agglomerated distribution focused on Middle Chile.	273; 276; 278
Compare and contrast the territorial shapes of Uruguay and Chile.	Uruguay's is compact, easy to govern and manage; Chile's is severely elongated, causing potential political problems.	276; 278
In what ways does Paraguay exhibit traits of a regional transition zone?	Paraguay is associated with both the Andean West and the Southern Cone. 95% of Paraguay's population is mestizo, with very strong Amerindian influences. Amerindian Guaraní is spoken alongside Spanish. Physiography is non-Andean, and Paraguayan economic geography is reorienting to the south.	277
What are northern Chile's major minerals?	Nitrates in the Atacama; copper, too, especially in the vicinity of Chuquicamata.	278

MAP EXERCISES

Map Comparison

1. The impact of the Andes on South America's physical and cultural geography is discussed on pp. 268-273. Compare the maps referred to, setting forth your observations in some detail. Reread the box on altitudinal zonation (p. 228), and discuss the application of this scheme to the Andes in the country of Peru.

2. South America's complex cultural geography often affects the ability of a state to integrate and effectively govern its citizens. After carefully studying Fig. 5-5 (p. 247), rank the realm's countries according to their apparent cultural uniformity—assuming that the more homogeneous cultures are the stronger nation-states. Provide a brief justification for each country, but qualify it if you have learned appropriate additional information that might affect overall unity (or lack thereof).

3. On Fig. 5-11 (text p. 269), rule a straight line in light pencil from Callao-Lima on the coast to Iquitos in the northeastern *Oriente*; draw an identical line on the physiographic map (Fig. 5-1, p. 238). On a blank piece of paper draw a cross-section of this traverse (using both maps to assist you), showing the general configuration of the surface involved (your instructor can assist you in doing this). Then, using Fig. 5-5 (p. 247), add in the cultural association of each segment of the cross-section, and describe its leading characteristics according to the text discussion (pp. 247-248).

Map Construction *(Use outline maps at the end of this chapter)*

1. In order to familiarize yourself with South American physical geography, place the following on the first outline map:

    a.   *Rivers*: Amazon, Orinoco, Paraguay, Paraná, Uruguay, São Francisco, Magdalena, Cauca, Apure, Marañon, Madeira, Guayas, Rio de la Plata, Colorado, Xingu, Tocantins, Negro

    b.   *Water bodies*: Amazon Delta, Lake Maracaibo, Lake Titicaca, Gulf of Guayaquil, Strait of Magellan, Plata Estuary, Peru (Humboldt) Current

    c.   *Land bodies and features*: Cape Horn, Tierra del Fuego, Falkland Islands, Chiloé Island, Guiana Highlands, Mato Grosso Planalto, Andean Altiplano, Patagonian Plateau, Pampa, Chaco, Llanos, Brazilian Highlands, Atacama Desert, Andes Mountains

2. On the second map, political-cultural information should be entered as follows:

    a.   Label each country, and its capital (*).

    b.   Reproduce the cultural map (Fig. 5-5, p. 247) using an appropriate color-pencil scheme.

3. On the third outline map, urban-economic information should be entered as follows:

    a.   *Cities* (locate and label with the symbol ●): Caracas, Ciudad Bolívar, Georgetown, Paramaribo, Cayenne, Valencia, Barquisimeto, Cartagena, Medellín, Bogotá, Buenaventura, Quito, Esmeraldas, Guayaquil, Lima, Callao, Iquitos, Huancayo, Arequipa, La Paz, Oruro, Potosí, Asunción, Arica, Antofagasta, Chuquicamata, Valparaíso, Santiago, Concepción, Valdívia, Montevideo, Buenos Aires, Rosario, Córdoba, Mendoza, Tucumán, Punta Arenas, São Paulo, Rio de Janeiro, Santos, Florianópolis, Pôrto Alegre, Belo Horizonte, Curitiba, Brasília, Volta Redonda, Salvador, Recife, Belém, Fortaleza, Manaus

b. *Economic regions* (identify with circled letter):

A - Itaipu Dam
B - Yacyretá Dam
C - Pampas
D - *Oriente*
E- Magdalena Valley
F - Guayas Lowland
G - Titicaca Basin
H - Middle Chile
I- Minas Gerais
J - Grande Carajás Scheme

PRACTICE EXAMINATION

Short-Answer Questions

Multiple-Choice

1. South America's largest city in population size is:

   a) Mexico City   b) São Paulo   c) Buenos Aires
   d) Rio de Janeiro   e) Caracas

2. The landlocked country whose search for an outlet to the sea was thwarted by Chile is:

   a) Bolivia   b) Suriname   c) Portuguese Guiana
   d) Paraguay   e) Amazonia

3. Which of the following regions is not located in Argentina?

   a) Patagonia   b) Pampa   c) Maracaibo Lowland
   d) Chaco   e) Entre Rios

4. Brazil does not border:

   a) Paraguay   b) Venezuela   c) Chile
   d) Peru   e) Argentina

5. Which of the following countries harbored an insurgent state at the opening of the twenty-first century?

   a) Uruguay   b) Paraguay   c) Brazil
   d) Colombia   e) Argentina

6. Which of the following countries is a member of Mercosur?

a) Argentina   b) Bolivia   c) Venezuela
d) Ecuador   e) Suriname

True-False

1. Suriname is a former colony of France, and today enjoys complete independence.

2. Uruguay is a classic example of an elongated state.

3. Itaipu Dam is situated on the border between Brazil and Chile.

4. Belo Horizonte and Salvador are Amazonian Basin growth poles.

5. Ecuador's largest city is also a capital.

6. Suriname is notone of the three Guianas.

Fill-Ins

1. The two leading resources of Chile's Atacama Desert are nitrates and _____.

2. At the center of the "Latin" American City model, one finds a land-use zone called the _____.

3. The coffee-growing areas of Colombia are concentrated in the altitudinal zone called *tierra* _____.

4. Brazil's largest city is _____.

5. Argentina's southernmost plateau region is called _____.

6. The country that borders Chile on its north is _____.

Matching Question on South American Countries

____1.	Still under colonial rule	A. Colombia
____2.	Former Dutch Guiana	B. Venezuela
____3.	Fought war for Falkland Islands	C. Guyana
____4.	Cuzco Basin	D. Suriname
____5.	Contains realm's largest city	E. French Guiana
____6.	East side of Plata estuary	F. Ecuador
____7.	Elongated state	G. Peru
____8.	Capital is La Paz	H. Bolivia
____9.	Lower Orinoco Basin	I. Paraguay
___10.	Population more than 50% Asian	J. Chile
___11.	Site of Cusiana oilfield	K. Argentina
___12.	Regional transition zone	L. Uruguay
___13.	Guayas Lowland	M. Brazil

Essay Questions

1. Despite internal social and economic problems, Brazil still possesses vast growth potential. Discuss the advantages of Brazil's developmental opportunities, emphasizing its natural resources in a subregion-by-subregion survey.

2. Discuss the altitudinal zonation scheme of Middle and South American environments (introduced in the box on text p. 228 in Chapter 4) as it applies to the Andean highlands of northwestern South America, highlighting the distinctive cultural and economic geographies associated with each zone.

3. Discuss the historical geography of Peru over the past 1000 years. Why did great advances occur here in the pre-European period? Why did the Spaniards headquarter their colonial empire here after 1535? Why has Peru ceased to be one of South America's leading countries in this century?

4. Discuss the agricultural geography of South America. Draw a sketch map showing the commercial, subsistence, and non-agricultural areas of the realm. Why did this particular pattern emerge, and what relationship does it bear to the cultural spatial differentiation of the continent?

5. The economic geography of northern and western South America has in part been shaped by the exploitation of major mineral resources. Discuss the spatial patterns of production that characterize the following metallic and fossil-fuel resources: oil and natural gas, nitrates, copper, silver, and iron ore. What key minerals are missing or are unfavorably distributed in the drive to modernize these countries?

TERM PAPER POINTERS

The "Term Paper Pointers" section of the Introduction chapter in this **Study Guide** offered suggestions about approaching research and writing on geographic realms and their components, and you may wish to consult this material if you are undertaking a report on a South American region.

You should also be aware of the textbook's Web Site, which includes direct links to a number of other Web Sites that may be quite helpful in your research. It is: http://www.wiley.com/college/regions2000

South America's diversity again necessitates that one obtain a deeper background from a good regional geography. The best work listed in the **References and Further Readings** section is James & Minkel, but this book is a shadow of its earlier editions; the 4th edition by James alone (1969) remains a classic, but it is outdated in most of the subjects it covers. Of the more recent works (Blakemore & Smith, Bromley & Bromley, Clalwson, Gilbert(1990), Morris, and Preston) none rank with James, but they are useful because of their currency and because they contain good bibliographies of the regional and systematic geographical literature of the past few years (an important feature for studying a realm that is undergoing steady change). Two additional overviews are Blouet & Blouet and Caviedes & Knapp, for the most part arranged topically rather than regionally; both contain general discussions of South American environments, historical geography, transportation, agriculture, population characteristics, urbanization, mining and manufacturing, and brief regional profiles. Brawer's atlas is a good guide to the continent, and much useful information is available in Collier.

Several cross-matching topical works are worth noting, too, many from disciplines outside geography. Cultural and historical geography are treated in Augelli, Crow, Gade, Greenfield, Hansis, Smith et al., Sponsel, and Tenenbaum. Political geography is covered by Barton, Caviedes, Kelly, Kent, McColl, Radcliffe & Westwood, Rohter, and St. John (both titles). Issues of economic development are highlighted in Becker & Egler, Box, "Brazil. . .," Brooke, Bromley & Bromley, Browder & Godfrey, Eakin, Goulding et al., "Growth in. . .," Gwynne, Keeling, Krauss, LaFranchi, Margolis, O'Connor, Page, and World Bank. Agriculture is treated in "Brazil. . .," Gade, "Growth in. . .," LaFranchi, Preston, Smith et al., and Sponsel. Urbanization trends are elucidated by Brunn & Williams, Ford, Fuchs, Gilbert (1996), Greenfield, Griffin & Ford (both titles), Gugler, Gwynne, Lowder, Nash, "Sao Paulo," Potter, and Wilkie. The geography of cocaine is treated in Clawson & Lee and Rengert. The Amazon environmental crisis is covered in Browder & Godfrey, Goulding et al., Margolis, Myers, O'Connor, Smith et al., and Sponsel.

SOUTH AMERICA

| 0 | 400 | 800 | 1200 | 1600 Kilometers |
| 0 | 200 | 400 | 600 | 800 | 1000 Miles |

Equator

0°

Tropic of Capricorn

20°

40°

80°

60°

40°

100°

80°

60°

40°

20°

0°

20°

40°

SOUTH AMERICA

| 0 | 400 | 800 | 1200 | 1600 Kilometers |
| 0 | 200 | 400 | 600 | 800 | 1000 Miles |

Equator

0°   0°

20°   20°

Tropic of Capricorn

40°   40°

80°   60°   40°

100°   80°   60°   40°   20°

SOUTH AMERICA

| 0 | 400 | 800 | 1200 | 1600 Kilometers |

| 0 | 200 | 400 | 600 | 800 | 1000 Miles |

Equator

Tropic of Capricorn

SOUTH AMERICA

			1600 Kilometers
0	400	800	1200
0	200	400	600	800	1000 Miles

Equator

0°

20°

Tropic of Capricorn

40°

80°

60°

40°

100°

80°

60°

40°

20°

CHAPTER 6

NORTH AFRICA / SOUTHWEST ASIA

OBJECTIVES OF THIS CHAPTER

Having completed our survey of the Western Hemisphere's realms, we now return to the Eastern Hemisphere, whose remaining Old World realms will occupy our attention for the rest of the book. Historically, the North Africa/Southwest Asia realm is properly regarded as the world culture hearth. Following a brief introduction and definitional discourse on the complexity of this realm, its historical evolution is traced emphasizing the prominence of religion, particularly Islam. Traditional culture, of course, stills weighs heavily on the daily affairs of this realm, and even its vast oil production is influenced by these considerations. An in-depth regional treatment follows, highlighting such major trouble spots as Israel, Iraq, Iran, Cyprus, Lebanon, the Horn of Africa, and the Persian Gulf. Turkestan is the realm's seventh region; this is former Soviet Central Asia, (now constituted by Kazakhstan, Turkmenistan, Uzbekistan, Kyrgyzstan, Tajikistan), to which we have added Afghanistan.

Having learned the regional geography of North Africa/Southwest Asia, you should be able to:

1. Appreciate the complexities involved in defining and naming this realm.

2. Understand the realm's basic cultural geography.

3. Describe the history of this realm, stressing its role in the development of many of the world's leading religions, particularly Islam.

4. Appreciate the significance of Islam for this realm as a whole, and the internal geographic variations of that faith.

5. Explain the major processes of spatial diffusion and be aware of the broad geographic patterns they shape.

6. Describe the production of oil in this realm, and the impact it has had on the development of countries that contain petroleum supplies.

7. Understand the major trends within each of this realm's regions, and why so many global political problems have arisen here.

8. Locate the major physical, cultural, and economic-spatial features of the realm on an outline map.

GLOSSARY

Qanats *(282)*

An underground tunnel built to carry irrigation water by gravity flow from nearby mountains (where orographic precipitation occurs) to the arid flatlands below.

Cultural geography *(285)*

The wide-ranging and comprehensive field of geography that studies spatial aspects of human cultures.

Culture hearth *(285)*

A source area or innovation center from which cultural traditions are transmitted.

Cultural diffusion *(285)*

The outward spreading of a culture trait from its hearth to other places.

Cultural ecology *(285-286)*

The multiple relationships between human cultures and their natural environments.

Mesopotamia *(286)*

The Tigris-Euphrates Plain of present-day Iraq—literally "land amidst the rivers"—which is the hearth of civilization.

Fertile Crescent *(286)*

An arc stretching from the eastern Mediterranean coast to near the Persian Gulf, site of early plant domestications and farming innovations (see Fig. 6-3).

Hydraulic civilization theory *(286)*

Civilizations able to control irrigated farming over large hinterlands; often held power over others in less fortuitous locations.

Climate change theory *(287)*

An alternative to the *hydraulic civilization theory*; holds that changing climate (rather than a monopoly over irrigation methods) could have provided certain cities in the ancient *Fertile Crescent* with advantages over others.

Spatial diffusion (289-box)

The spatial spreading or dissemination of a phenomenon across space and through time.

Expansion diffusion (289-box)

The spreading of an idea or innovation through a fixed population in such a way that the number of those adopting grows continuously larger.

Relocation diffusion (289-box)

Diffusion by migration wherein innovations are carried by a relocating population.

Contagious diffusion (289-box)

Local-scale diffusion, strongly controlled by distance from the point of origin.

Hierarchical diffusion (289-box)

Macro-scale diffusion through a national or continental-scale urban hierarchy, involving the "trickling down" of an innovation from atop a hierarchy to each of the lower levels in turn.

Culture region (291)

A distinct, culturally discrete spatial unit; a region within which certain cultural norms prevail.

Religious fundamentalism (292)

Religious movement whose objectives are to return to the foundations of that faith and to influence state policy.

Organization of Petroleum Exporting Countries (OPEC) (296-297-map;298)

The international cartel constituted by eleven member states; the North Africa/Southwest Asia OPEC countries are mapped in Fig. 6-8.

Cultural revival (299)

The regeneration of a long-dormant culture through internal renewal and external infusion.

Basin irrigation (302)

An ancient irrigation method of the lower Nile Valley involving the trapping and later release of floodwaters.

Perennial irrigation (302)

The more modern Egyptian irrigation technique, using dams and levees to store and regulate the use of floodwater throughout the year.

Fellaheen (302)

Egypt's peasant farmers, who still struggle to eke out a subsistence level of existence.

Maghreb (305)

The western region of this realm, consisting of the northwesternmost African countries of Morocco, Algeria, and Tunisia. The name itself means "Isle of the West."

Tell (306)

The lower slopes and coastal plains of northwesternmost Africa between the Atlas Mountains and the sea.

Sahel (307)

Arabic for "border," refers to an east-west belt stretching across the southern edge of the Sahara (the heart of the African Transition Zone).

Muslim Front (308)

A term used by certain scholars for the African Transition Zone of northern Africa, which is primarily regarded as a still-expanding frontier of Islam that affects countries from Guinea in the west to the African Horn in the east.

Stateless nation (310)

A national group that aspires to become a nation-state but lacks the territorial means to do so; the Palestinians and Kurds of Southwest Asia are classic examples.

Nomadism (326)

A way of life pursued by people who migrate cyclically among a set of places, usually practicing pastoralism.

Buffer state (332)

Part of a *buffer zone*–a set of countries separating ideological or political adversaries. In southern Asia, Afghanistan, Nepal, and Bhutan were parts of a buffer zone between British and Russian-Chinese imperial spheres. Thailand was a *buffer state* between British and French colonial domains in mainland Southeast Asia.

SELF-TESTING QUESTIONS

Cover the right side of the page with a sheet of paper. Uncover each line after you have attempted to answer the question in the left column. If necessary, refer to the textbook page(s) listed at the right.

Question	Answer	Page
Realm Characteristics		
Why is "Arab World" a misleading title for this realm?	This is a *linguistic* term which does not fit much of the realm.	282
Why is "Islamic World" just as unsatisfactory?	The Muslim religion prevails far beyond the realm, and within it there are a number of countries in which Islam is not the dominant faith.	283
Why is "Middle East" not much of an improvement?	It is imprecise and reflects a Western perceptual bias—but it is based on multiple criteria (unlike the two preceding labels).	283-284
Cultural Geography		
What is cultural geography?	The wide-ranging and comprehensive field that studies spatial aspects human cultures, focusing on not only culture landscapes but also culture hearths.	285
What are the goals of those who specialize in cultural diffusion?	To be able to reconstruct ancient routes by which knowledge and achievements of culture hearths spread (diffused) to other areas.	285
Historical Geography		
What were the three culture hearths of North Africa/ Southwest Asia?	The Tigris-Euphrates Plain (Mesopotamia), the lower Nile Valley, and the peripheral lower Indus valley (which in this book is treated as part of the South Asian realm).	286-287
Name some of the major agricultural innovations that originated in these hearths.	Wheat, rye, barley, peas, beans, grapes, apples, peaches, horses, pigs, and sheep.	287

When did the Prophet Muhammad live?	A.D. 571-632; Muhammad became a religious leader in 613.	288
What are the "Five Pillars" of Islam?	Repeated statement of the creed; daily prayer; a month of daytime fasting; almsgiving; and a pilgrimage to Mecca.	289

Spatial Diffusion

What is an *innovation wave*?	The stages of Hägerstrand's model, showing the progress of a diffusing phenomenon, which pulses outward from its hearth across the adopting region.	289-box
What are the two major types of spatial diffusion?	*Relocation diffusion* (diffusion by migration) and *expansion diffusion* (diffusion within a fixed population).	289-box
What are the two forms of expansion diffusion?	*Contagious diffusion* (distance-controlled local diffusion) and *hierarchical diffusion* (coursing downward within a national urban hierarchy).	289-box

Oil Resources

What proportion of the world's oil reserves are contained within this realm?	More than 65 percent.	295
What is OPEC?	The Organization of Petroleum Exporting Countries—the international oil cartel or syndicate formed by producing countries to promote their common economic interests through the formulation of joint pricing policies and the limitation of market options for consumers.	298

Egypt and the Lower Nile Basin

How important is the Nile to this country?	This river is Egypt's lifeline; valley and delta are home to about 95% of the population.	302
How do *basin* and *perennial* irrigation differ?	The former is an ancient method of trapping floodwaters; the latter is a modern system based on dams, of which the Aswan High Dam is most prominent.	302

Why is Egypt so prominent an Arab country?	Its pivotal location, maintained throughout history; it lies at the crossroads where North Africa and Southwest Asia come together—the heart of the Arab World.	302-303
Name Egypt's six major subregions.	Nile Delta; Middle Egypt (including Cairo); Upper Egypt; Western Desert; Eastern Desert/Red Sea Coast; Sinai Peninsula. See Fig. 6-10.	303-map; 305

The Maghreb and Libya

What does Maghreb mean? Name its constituent countries.	Literally "Isle of the West," based on the Atlas Mountains standing above the flat Sahara. Morocco, Algeria, and Tunisia; neighboring Libya is often discussed together with the Maghreb.	305
In what way is the political landscape of the Maghreb changing?	Morocco, Algeria, and Tunisia face Islamic fundamentalism, partly due to frustration from weakened economies.	306

African Transition Zone

What is the extent of this region?	The east-west band of states along the southern margin of the Sahara, from Mauritania and Senegal in the west to Ethiopia and Somalia in the eastern African Horn. See Fig. 6-12.	307
Do the countries of coastal West Africa contain as large a Muslim population as in North Africa?	No; while the countries of North Africa are over 90 percent Muslim, the nations of coastal West Africa are well below 50 percent Muslim. (See Fig. 6-12 on p. 307).	307
Why do some observers refer to the African Transition Zone as the Muslim Front?	This is a zone where Islam continues to expand–a religious frontier affecting two dozen African countries.	308

The Middle East

Name the 5 countries of this region.	Iraq, Syria, Jordan, Lebanon, and Israel.	308

What are the leading problems of Iraq, Syria, and Jordan?	Iraq is bedeviled by government corruption and a brutal dictator whose military adventures in neighboring countries have overwhelmed the (surprisingly good) chances for economic development; Syria continues to negotiate with Israel for the return of the Golan Heights; Jordan is beset by a rapidly growing population in which refugees outnumber natives, and filling the leadership void left by its late King Hussein.	308-311
Why has Lebanon been torn by internal strife for the past two decades?	The old political accommodation between the Muslims and Christians no longer fits a new reality in which Muslim and other non-Christian groups (themselves disunified) have come to dominate the population.	311-312
How old is the modern state of Israel?	It was founded amid much turmoil in 1948, and has lived with threatening neighbors ever since.	312
What is the distribution of Palestinians in the Middle East?	At the beginning of the twenty-first century, nearly 8.5 million are dispersed throughout the region; Jordan contains over 2.4 million, and Syria and Lebanon contain sizeable concentrations as well; Israel and its occupied territories are home to more than 4.1 million.	315-box
How has Israel's development fared vis-à-vis its neighbors?	Far better; while the Arab countries remain relatively unproductive and mired in poverty, Israeli prosperity and economic growth has been impressive.	317
What percentage of the Israeli population is urbanized?	Just over 90 percent.	317

Arabian Peninsula

Name the countries of this region.	Saudi Arabia, Kuwait, Bahrain, Qatar, the United Arab Emirates, the Sultanate of Oman, and the Republic of Yemen.	318
What are some major recent Saudi Arabian development projects?	The construction of an effective internal transportation and communications network; the growth of the petrochemical and metal industries; the new cities of Jubail and Yanbu.	318-319

The Empire States

Name the countries of this region.	Turkey, Iran, and the northern part of Cyprus.	321
Why has Turkey remained aloof from other Arab countries?	Greater European influences shaped culture here; Atatürk's revolution stressed the detachment from other Islamic states, and that tradition endures.	321-322
Who are the contestants in the Cyprus struggle?	Greeks who seek union with Greece; Turks who want ties to Turkey.	325-box
How has modernization affected Iran?	Surprisingly little; it was also a contributory factor to the 1979 revolution that deposed the Shah.	326
How did the Iran-Iraq War (1980-1990) affect Iran?	The ten-year struggle left Iran poorer and weaker: oil revenues were spent on unproductive pursuits.	326

Turkestan

Which countries comprise Turkestan?	Kazakhstan, Turkmenistan, Uzbekistan, Kyrgyzstan, Tajikistan, and Afghanistan. Before 1992, these countries constituted the former Soviet Union's Central Asian republics (with the exception of Afghanistan).	328
Why is Kazakhstan's relative location important?	It forms a corridor between the Caspian oil reserves and China (a large oil consumer). Future pipelines could be crucial to China.	329
What ethnic minorities are located in Turkmenistan?	Russians, Uzbeks, Kazakhs, Ukrainians, Tatars, and Armenians.	329
In what way are political relations troubled for Uzbekistan?	Relationships with neighbors are tense—irredentism toward Uzbek minorities in other nations is on the rise; internal strife with non-Uzbeks continues.	331
Which states established the buffer role for Afghanistan?	19th century Britain and Russia, for which the country was a neutral zone lying between their Asian spheres of influence.	332
How has Turkestan fared since the end of Soviet rule?	Government authoritarianism has continued— corruption, inefficiency, and instability plague this region; Islamic fundamentalism is on the rise; and the area suffers from ethnic and cultural strife.	333

MAP EXERCISES

Map Comparison

1. Compare the map of world religions (Fig. 6-2, pp. 284-285) to the map of world geographic realms (pp.4-5). Which realms are clearly identifiable with a major religion? Which realms exhibit multiple religions that suggest internal cultural conflicts? Double-check your observations against the world population map (pp. 20-21), underscoring those realms associated with large population clusters.

2. Prepare a brief essay on the distribution of petroleum reserves in the North Africa/Southwest Asia realm based on your analysis of Fig. 6-8 (pp. 296-297). When possible, compare to patterns of production as shown in the bar graph at the bottom of the map.

3. Compare the map of Turkestan (Fig. 6-19, p. 327) to the following maps in Chapter 2: Fig. 2-5 (p. 120), Fig. 2-6 (pp. 124-125), and Fig. 2-7 (pp. 126-127). How well does Turkestan "fit" the North Africa/Southwest Asia realm, a question that also needs to be weighed in the context of the map of world religions (Fig. 6-2, pp. 284-285)?

Map Construction   (Use outline maps at the end of this chapter)

1. In order to familiarize yourself with North Africa/Southwest Asia physical geography, place the following on the first outline map:

    a.   *Rivers*: Nile, White Nile, Blue Nile, Tigris, Euphrates, Orontes, Jordan, Shatt al Arab, Amu Darya, Syr Darya

    b.   *Water bodies*: Red Sea, Persian Gulf, Suez Canal, Gulf of Aqaba, Gulf of Suez, Bab el Mandeb Strait, Hormuz Strait, Strait of Gibraltar, Gulf of Sidra, Caspian Sea, Aral Sea, Black Sea, Mediterranean Sea, Lake Nasser, Lake Tana, the Sudd, Dead Sea, Bosporus Strait, Gulf of Aden, Gulf of Oman, Arabian Sea, Garagum Canal, and the new Mesopotamian Drainage Canal

    c.   *Land bodies*: Sinai Peninsula, Cyprus, Horn of Africa, Socotra, Bahrain, Canary Islands, Kyrgyz (Kirghiz) Steppe

    d.   *Mountains and Deserts*: Atlas Mountains, Tibesti Mountains, Libyan Desert, Sahara Desert, Arabian Desert, Rub al Khali (Empty Quarter), Anatolian Plateau, Ahaggar Mountains, Ethiopian Highlands, Hejaz Mountains, Negev Desert, Plateau of Iran, Elburz Mountains, Zagros Mountains, Pamir Mountains, Tian Shan Mountains, Hindu Kush Mountains, Nubian Desert, An Nafud Desert

2. On the second map political-cultural information should be entered as follows:

   a. Label each country, and label and locate each capital city with the symbol *.

   b. Color each country according to its major religion, using information and appropriate color symbols from the map on text pp. 284-285 (be sure to differentiate between the Sunni and Shi'ite sects in Muslim states).

3. On the third outline map, economic-urban information should be entered as follows:

   a. *Cities* (locate and label with the symbol ● /capitals should again be shown with symbol *): Casablanca, Marrakech, Rabat, Tangiers, Oran, Algiers, Tunis, Tripoli, Benghazi, N'Djamena, Niamey, Bamako, Timbuktu (Tombouctou), Nouakchott, Port Sudan, Khartoum, Asmara, Djibouti, Mogadishu, Cairo, Alexandria, Port Said, Ismailia, Aswan, Riyadh, Mecca, Medina, Dhahran, Jubail, Yanbu, Kuwait City, Manama, Doha, Abu Dhabi, Muscat, Aden, San'a, Amman, Jerusalem, Haifa, Tel Aviv-Jaffa, Eilat, Aqaba, Beirut, Sidon, Damascus, Aleppo, Nicosia, Baghdad, Basra, Kirkuk, Mosul, Istanbul, Ankara, Izmir, Adana, Zonguldak, Diyarbakir, Tehran, Tabriz, Abadan, Mashhad, Shiraz, Kabul, Herat, Baki, Ashgabat, Mary, Toshkent, Nukus, Dushanbe, Bishkek, Almaty, Qaraghandy, Astana

   b. *Economic regions* (identify with circled letter):

      A - African Horn
      B - West Bank
      C - Mesopotamia
      D - Lower Egypt
      E - Gaza Strip
      F - Wakhan Corridor
      G - Asia Minor
      H - The Tell
      I - Persian Gulf oilfield
      J - Jazirah
      K - Tengiz Basin
      L - Russian Transition Zone

PRACTICE EXAMINATION

Short-Answer Questions

Multiple-Choice

1. Which of the following countries is part of the Maghreb region?

   a) Libya   b) Egypt   c) Algeria
   d) Turkmenistan   e) Turkey

2. Which of the following rivers flows through Iraq?

   a) Euphrates   b) Jordan   c) Blue Nile
   d) White Nile   e) Aral

3. Kemal Atatürk is most closely identified with the city of:

   a) Diyarbakir   b) Mecca   c) Tehran
   d) Ankara   e) Baghdad

4. The Taliban revolution of the mid-1990s occurred in:

   a) Kazakhstan   b) Afghanistan   c) Egypt
   d) Cyprus   e) Azerbaijan

5. The Ottoman Empire was centered in the modern-day country of:

   a) Palestine   b) Egypt   c) Turkey
   d) Greece   e) Russia

6. Which of the following countries is located in the African Horn?

   a) Syria   b) Senegal   c) Yemen
   d) Sudan   e) Somalia

True-False

1. The West Bank region is hotly contested by Israel and Syria today.

2. The region known as the Sahel is located in the African Transition Zone.

3. Hierarchical diffusion involves the spreading of an innovation across a large area, "trickling- down" within a country's overall urban system.

114

4. Both Iran and Iraq are dominated by Shi'ite Muslims.

5. Kazakhstan's new capital is named Astana.

6. The Persian Gulf and Mediterranean Sea are connected by the Suez Canal.

Fill-Ins

1. The eastern Mediterranean island of _____ is hotly contested by Greece and Turkey.

2. _____ diffusion depends upon a migrating population to transmit an innovation.

3. The _____ Muslim sect dominates life in Iran today.

4. The West Bank is so named with respect to the _____ River.

5. The capital of Egypt is _____.

6. The mountains located between the Black and Caspian Seas are named the _____.

Matching Question on the Realm's Countries

____1.	Home of the *fellaheen*	A. Israel
____2.	Maghreb kingdom	B. Syria
____3.	North Africa's most radical state	C. Uzbekistan
____4.	Headquarters of Ottoman Empire	D. Turkey
____5.	Addis Ababa	E. Libya
____6.	Garagum Canal	F. Saudi Arabia
____7.	Jubail planned city	G. Kazakhstan
____8.	Shi'ite sect dominant	H. Algeria
____9.	Occupies the heart of Turkestan	I. Iraq
____10.	Invader of Kuwait in 1990	J. Jordan
____11.	Zionist movement	K. Egypt
____12.	Tell Atlas	L. Morocco
____13.	Tengiz Basin oilfield	M. Turkmenistan
____14.	Refugees outnumber natives	N. Iran
____15.	Lost Golan Heights to Israel	O. Ethiopia

Essay Questions

1. Geographic definitions of this realm have been difficult to arrive at and are still not entirely satisfactory. Compare and contrast the "Arab World" and "Islamic World" definitions, highlighting the strengths and weaknesses of each. Does the label "Middle East" add anything useful to the argument, and why is it not used in the text as the realm's name?

2. The influence of the Muslim faith has been a dominant shaping force in the history of this realm for the past 1400 years. Elaborate on the diffusion and acceptance of Islam in North Africa/ Southwest Asia since A.D. 600, highlighting cultural-geographical similarities and differences.

3. Discuss the regional problems of the Maghreb, highlighting environmental variables, internal political differences, resource distributions, and economic development opportunities.

4. Discuss the changing political geography of Israel since its creation in 1948. How has this country strengthened itself since the 1967 War, and what is the present likelihood of a more stable coexistence with its Arab neighbors?

5. Compare and contrast the political and economic experiences of Saudi Arabia and Iran in the past ten years, and evaluate the chances for meaningful modernization and development in the two countries during the coming few years.

TERM PAPER POINTERS

The "Term Paper Pointers" section of the Introduction chapter in this **Study Guide** offered suggestions about approaching research and writing on geographic realms and their components, and you may wish to consult this material if you are undertaking a report on a North African or Southwest Asian region. You should also be aware of the textbook's Web Site (http://www.wiley.com/college/regions2000), which includes direct links to a number of other Web Sites that may be quite helpful in your research.

The best approach to this sprawling realm is again by consulting a basic regional geography. Two recent titles are listed in the **References and Further Readings** section and should be consulted first—Held and Beaumont et al. Both titles by Chapman & Baker are useful too, but their surveys are rather brief; the rest of this regional overview literature is dated but still adequate, though somewhat dominated by a British point of view—Fisher, Longrigg, and Cressey. The recently-added region of Turkestan (former Soviet Central Asia), is covered by "A Caspian Gamble. . . ," Adshead, Chinn & Kaiser, Fuller, Gleason, and Lewis. Atlases worth looking at are those by Blake et al., Brawer, Freeman-Grenville, Gilbert (1994), Lemarchand, and Robinson. Other regions are covered in Gilbert (1998), Gradus & Lipshitz, Gurdon, Joffe, Matvejevic, Meiselas, "New and Old," "Saudi Arabia's Centennial," and Zoubir. Additional overviews of the realm can be obtained from Findlay, Kemp & Harkavy, Hourani, Lamb, Mostyn, Murphey, Rahman, and Sluglett & Farouk-Sluglett; the dictionary by Hiro is a particularly helpful reference.

Numerous topical treatments of the realm's geography are available as well. Economic geography and development is found in Rahman, "Saudi Arabia's Centennial," and Zoubir. The geography of arid lands, and the closely related topic of scarce water resources, is treated by Beaumont, Cloudsley-Thompson, Heathcote, Kliot, Soffer, and Swearingen & Bencherifa. The Arab-Israeli conflict and its background is covered in Dowry, Fromkin, Hourani, Lamb, Newman, and Peters. Politico-geographical issues are surveyed in Azarya, Cohen, Drysdale & Blake, Joffe, Kemp & Harkavy, Meiselas, Newman, Prescott, and Soffer. Population geography is examined in Goldscheider, Lewis, and Omran & Roudi. Urban geography is covered in Amirahmadi & El-Shaks and Rodenbeck. Historical geography is treated in Adshead, Dowry, Freeman-Grenville, Fromkin, Hourani, Meiselas, Murphey, Peters, and Robinson (1996). Aspects of the geography of religion, particularly Islam, are dealt with in Hourani, Lamb, Naipaul, Park, Rahman, Robinson (both titles), Smart, and Waines.

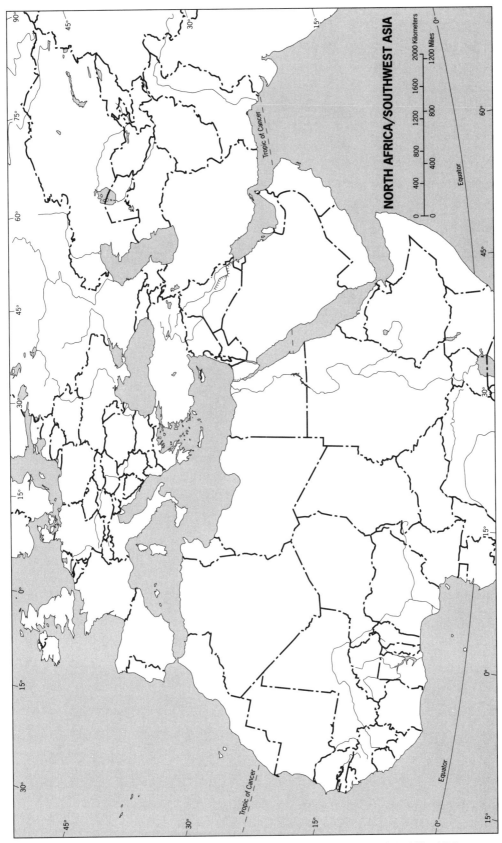

NORTH AFRICA/SOUTHWEST ASIA

Tropic of Cancer

Equator

Tropic of Cancer

Equator

2000 Kilometers
1200 Miles

1600
800

1200
800

800
400

400

0
0

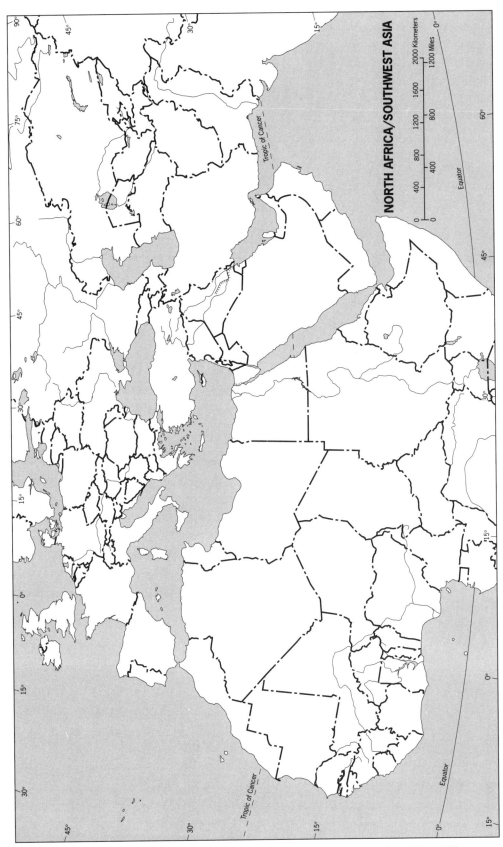

NORTH AFRICA/SOUTHWEST ASIA

Tropic of Cancer

Tropic of Cancer

Equator

0   400   800   1200   1600   2000 Kilometers
0   400   800   1200 Miles

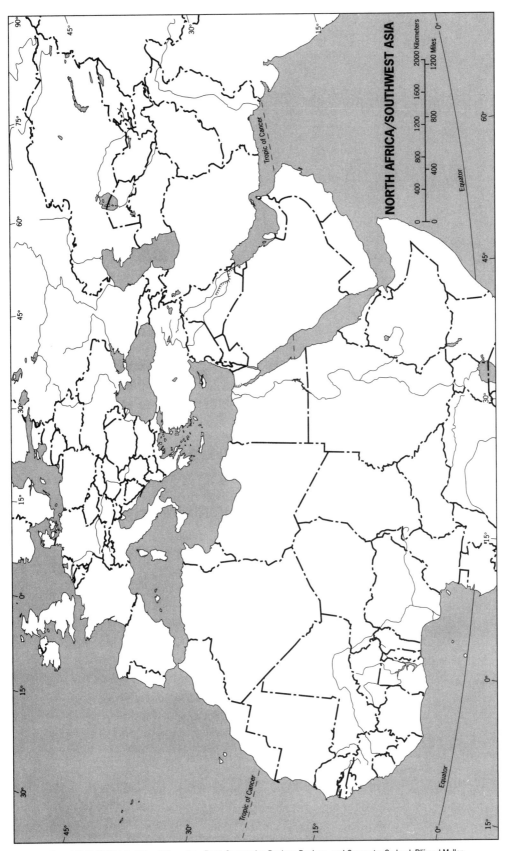

NORTH AFRICA/SOUTHWEST ASIA

Tropic of Cancer

Equator

Tropic of Cancer

Equator

2000 Kilometers

1200 Miles

0   400   800   1200   1600

0   400   800

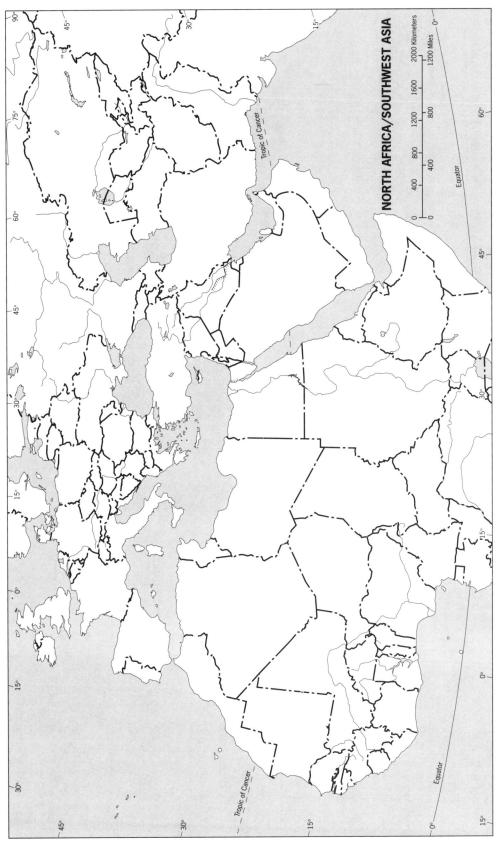

NORTH AFRICA/SOUTHWEST ASIA

Tropic of Cancer

Tropic of Cancer

Equator

Equator

0   400   800   1200 1600 2000 Kilometers
0   400   800   1200 Miles

CHAPTER 7
SUBSAHARAN AFRICA

OBJECTIVES OF THIS CHAPTER

Chapter 7 treats Subsaharan Africa, a problem-bedeviled realm in which life for the masses is almost always difficult. Following the introduction, Africa's environmental base is covered and highlights physiography, disease patterns, and the agricultural potential as well as risks. Historical geography comes next, examining pre-European cultures, the colonial transformation after 1450, and the post-independence era that dates from the late 1950s. The regional survey covers the rest of the chapter, focusing in turn on West, Equatorial, East, and Southern Africa.

Having learned the regional geography of Subsaharan Africa, you should be able to:

1. Understand the interpretation of African physiography and hydrography that is offered by the continental drift hypothesis.

2. Appreciate the concerns of contemporary medical geography and their applications to Africa.

3. Understand the spatial distributions of Africa's major endemic diseases and their impacts on daily life.

4. Grasp the importance of agriculture in African economic life, and the severe environmental risks that farming and herding are exposed to.

5. Understand the course of Subsaharan African history from the indigenous cultures through the colonial and post-independence eras.

6. Explain the political territorialization of modern Africa as a legacy of the colonial era that has finally ended.

7. Understand the cultural and economic trends that have shaped the regional structuring of West, Equatorial, East, and Southern Africa.

8. Understand the present politico-geographical situation in South Africa, including the racial polarization that intensified under the now-dismantled *apartheid* system, and the country's subsequent shift to democracy.

9. Locate the leading physical, cultural, and economic-spatial features of the realm on an outline map.

GLOSSARY

Rift valleys (336-337)

Geologic trenches formed when huge parallel cracks or *faults* occur in the Earth's crust, causing in-between strips of land to sink and form extensive linear valleys.

Great Escarpment (338)

The edge of the African plateau that forms a steep cliff, hundreds of miles in length, stretching north from southern South Africa to Tanzania on the east coast and as far as The Congo on the Atlantic coast.

Continental drift (338-339)

The hypothesis underlying the formation of Africa's physiography and hydrography. The breakup of the supercontinent called Pangaea and the subsequent drifting apart of its landmass components; Africa occupied the heart of Gondwana (Pangaea's southern part), and its physical geography is still shaped by this geologic functioning as the core-shield.

Medical geography (341)

The subdiscipline concerned with the spatial aspects of health and illness.

Endemism (341)

Refers to a disease in a host population, that affects many people in a kind of equilibrium without causing rapid and widespread deaths.

Epidemic (341)

A local or regional outbreak of a disease.

Pandemic (342)

An outbreak of a disease that spreads worldwide.

Land Tenure (343)

The way that people own, occupy, and use land.

Land Alienation (343)

Expropriation of (often the best) land by a conquering group.

Green Revolution (345-box)

The successful recent development of higher-yield, fast-growing varieties of rice and other cereals in certain developing countries. This led to increased production per unit area and a temporary narrowing of the gap between population growth and food needs; today, unfortunately, that gap is widening again.

Dhows (346;349)

Wooden boats with triangular sails, plying the seas of the Indian Ocean between the Arabian and the East African coasts.

Colonialsim (350)

The drive toward the creation and expansion of a colonial empire and, once established, its perpetuation.

Berlin Conference (350-box)

The 1884-1885 international conclave, convened by Germany's Chancellor von Bismarck, which superimposed on Africa a system of political boundaries agreed to by all colonial powers; amazingly, the scheme has survived for over a century and this political fragment-ation, for the most part, still underpins the realm's contemporary politico-geographical structuring.

Periodic market (362)

Village market that opens once every few days; part of a regional network of similar markets in rural, preindustrial societies where goods are brought to market on foot and barter remains a major mode of exchange.

Exclave (365)

A bounded (non-island) piece of territory that is part of a particular state but lies separated from it by the territory of another state.

Landlocked state (368)

A state with no outlet to the sea. Examples in this realm are Ethiopia, Burkina Faso, Chad, Uganda, and Zimbabwe.

Apartheid (374-375)

Literally, apartness. The Afrikaans term for South Africa's pre-1994 policies of racial separation, a system that produced highly segregated socio-geographical patterns.

Separate Development *(374)*

The spatial expression of South Africa's "grand" apartheid scheme, whereby nonwhite groups were required to settle in segregated "homelands." The policy was dismantled when white-minority rule collapsed in the early 1990s.

Highveld *(375)*

Dutch term for the high plateau of the South African interior.

Afrikaners *(375)*

The descendants of the original Dutch colonists of South Africa, known as the Boers, who steadily gained numerical and political strength until they came to power as the white-minority government from 1948-1994.

Coloured people *(376-377)*

Term for the racially mixed population resulting from the intermarriage of European settlers and Africans in South Africa's Cape area. Today, this mixed ancestry group forms the majority population in and around Cape Town.

SELF-TESTING QUESTIONS

Cover the right side of the page with a sheet of paper. Uncover each line after you have attempted to answer the question in the left column. If necessary, refer to textbook page(s) listed at the right.

Question	Answer	Page
African Environments		
What is a rift valley?	A trench formed when cracks or *faults* occur in the Earth's crust, and in-between strips sink down to create linear valleys	336-337
What is the *Great Escarpment?*	The margin of Africa's plateau, especially in Southern Africa, where the land surface drops sharply to the generally-narrow coastal lowland.	338
What is *continental drift*? How does it help us to understand Africa's physiography?	The hypothesis that all continents were once joined into a single supercontinent (Pangaea); Africa was located centrally in the southern component (Gond-wana); the giant landmass began to break apart at least 200 million years ago, with its continental components "drifting" away toward their present locations. Africa's unusual topographic and drainage patterns are most readily explained by this hypothesis.	338-339

| Describe Africa's broad patterns of climate and vegetation. | They are symmetrically distributed about the equator, which bisects the continent. The humid tropics are largely restricted to the western half since East Africa is mainly an elevated plateau; as one moves poleward from the tropics, steppes, significant deserts, and narrow strips of Mediterranean climates successfully prevail. | 339-341 |

Medical Geography

What is *medical geography*?	The study of health and disease within a geographic context and from a spatial perspective. Among other things, this field of geography examines the sources, diffusion routes, and distributions of diseases.	341
What are disease *vectors* and *hosts*?	Vectors are the intermediate transmitters (carriers) of pathogens; hosts are infected people.	341
What is the difference between an *epidemic* and a *pandemic* disease?	*Epidemics* are regional phenomena; an outbreak of global proportions is described as a *pandemic*.	341
How is sleeping sickness transmitted?	By the tsetse fly, native to most of tropical Africa, which infects both humans and livestock. See Fig 7-4.	341-342
What is Africa's worst pandemic disease?	Africa and the world's deadliest vectored disease is Malaria, whose carrier is a mosquito that prevails in most of Africa's inhabited areas.	342-343

African Agriculture

What problems plague African farming?	Poor tropical soils, excessive droughts, soil exhaustion, inadequate equipment, lack of capital, and inefficient methods.	343
What percentage of Subsaharan Africans depend on farming for their livelihood?	Approximately 75 percent.	344
What is the prospect for agricultural improvement?	Bleak for the foreseeable future; in the postcolonial period, food production has stagnated and at times declined; the life of the farmer remains difficult and gains in African agricultural output overall are more than offset by the rapid rate of population growth.	344-345

Historical Geography

Why is so little known about Subsaharan Africa before A.D. 1500?	The absence of written histories is the main cause, exacerbated by neglect and destruction of traditions and artifacts following the onset of colonialism and the disruptions of the slave trade.	345
Does this mean that pre-colonial Africa had no organized cultures and traditions?	Definitely not; on the contrary, sophisticated artifacts throughout Africa suggest advanced societies, wide-spread trade, and rich cultural legacies.	345-346
Where did the first European contacts occur?	In the 15th century, Europe's ships began to ply the waters of the entire west coast of Africa as the route to southern Asia opened. Way-station ports opened, and soon African middlemen began to organize trading of slaves, minerals, ivory, and spices.	348
How did the subsequent European penetration proceed?	Slowly at first, but after 1800 European powers finally laid claim to all of Africa's territory; expansion of spheres of influence soon caused disagreements but the 1884-85 Berlin Conference solved things peaceably. See box p. 350.	348-350; 350-box
What administrative systems did the Europeans employ in colonial Africa?	Indirect rule of colonies, protectorates, and territor-ies (Britain); paternalism (Belgium); assimilation toward the mother country (France); rigid economic control (Portugal).	350

Cultural Patterns

How many languages are spoken in this realm?	Over one thousand—dozens of languages in single countries, hundreds in a single region—make up the intricate linguistic jigsaw of Subsaharan Africa.	352
Which religion was brought to Subsaharan Africa with the first great wave of proselytism?	Islam.	354

West Africa

Which countries are included in this region?	Mauritania, Mali, Niger, Senegal, Gambia, Guinea-Bissau, Guinea, Sierra Leone, Liberia, Burkina Faso, Ivory Coast, Ghana, Togo, Benin, and Nigeria. Chad could be included too.	358-359; 359 - map

What regional unities exist?	A remarkable cultural-historical momentum; parallel east-west environmental belts, with much interaction perpendicular to them; an early economic orientation to world commodity markets.	358
Which part of West Africa contains most of its population?	The southern half of the region.	359
Who are Nigeria's three leading peoples?	The Yoruba of the southwest, the Ibos of the southeast, and the Hausa-Fulani of the semiarid north.	360
Which was the first of West Africa's states to become independent?	Ghana, in 1957.	362
What are *periodic markets*?	Local village markets that operate at regular intervals every few days; they attract nearby buyers and sellers.	362

Equatorial Africa

Which countries make up this region?	The Congo (former Zäire), Cameroon, Central African Republic, Congo, Gabon, Equatorial Guinea, southern Chad, and the southern subregion of Sudan. See Fig. 7-13.	363-364; 363-map
What has hampered The Congo's great development potential?	Lack of national cohesion, corrupt rule, and infrastructure problems.	364
What characteristics do the coastal countries of the Equatorial region share?	All four possess oil reserves and contain equatorial forests–oil and timber are exports.	364

East Africa

Describe East Africa's natural environment.	Plateau savanna country that becomes a steppe in the drier northeast; volcanoes and rift valleys mark the surface; Lake Victoria is a key feature.	365
Which countries constitute this region?	Tanzania, Kenya, Uganda, Rwanda, Burundi, and highland (southwestern) Ethiopia.	365
How does Kenya's development differ spatially from Tanzania's?	Kenya's development is concentrated in the core area around Nairobi; Tanzania's is much more dispersed.	366

| What has Tanzania's course been since independence? | It embarked on a socialist development program, but without adequate planning, so results were disappointing. A market-oriented "recovery" is now being employed, but the country remains one of the poorest in the world. | 366-367 |
| What is Uganda's current status? | The difficult struggle to emerge from the chaotic Amin years of the 1970s still continues. Economic recovery has been painstakingly slow; lately, the AIDS pandemic has inflicted a fearsome toll on this calamity-prone country. Conflict continues with Sudan. | 368 |

Southern Africa

Which countries are included in this region?	South Africa, Lesotho, Swaziland, Moçambique, Zimbabwe, Zambia, Malawi, Angola, Namibia, and Botswana (Madagascar [p. 371-box] is technically included as well). See Fig. 7-15.	370-map
What is the significance of the Copperbelt?	This zone of copper and other rich mineral deposits in Zambia's portion of Africa's greatest resource region—which stretches south from the southeastern portion of the The Congo through Zambia, Zimbabwe's Great Dyke, and into the heart of northern South Africa.	369-371
How have Moçambique's good chances for development been crushed?	A rebel movement severely damaged port facilities, hydroelectric projects, and plantations, and destroyed the tourist industry. It will take generations to rebuild the country.	372
What is the status of Namibia?	It became independent in 1990 after more than 70 years of South African administration; in 1994, the South African exclave of Walvis Bay was absorbed into Namibia.	374

South Africa's Characteristics

| What are the country's general geographic dimensions? | It is large and varied in its natural environments, and contains over 40 million people—the dominant state in Southern Africa. | 374 |
| Who were the Boers? | The descendants of the original Dutch settlers, who were driven from the Cape area by British colonists, but who eventually emerged in the north as the *Afrikaners* to take control as the white-minority government from 1948 to 1994. | 374-375 |

| Name the leading ethnic groups in South Africa today. | Besides the Afrikaners and British-heritage whites, there is a large number of black peoples (including the Zulu, Xhosa, and Sotho), the Cape Coloured, and the Indians (South Asians) of Natal Province. | 375; 375-table |

South African Political Geography

What is *apartheid*? What is its successor, *separate development*?	South Africa's unforgiving racial subjugation and segregation policy that after 1948 officially separated non-whites from whites; in the 1970s, the policy came to be called *separate development* and sought to isolate each racial group in its own "homeland" (in the process giving majority blacks the worst land and poorest resources). The dismantling of these systems took place during the mid-1990s.	375-376
Why was the world's attention focused on South Africa for the past decade?	The black opposition (spurred on by the African National Congress) began a major struggle; the "homeland" creation program began to fail; Nelson Mandela's 1990 release from prison and his subsequent activities were widely covered by global television; in 1994 Mandela was elected president in a democratic election, which triggered momentous change for South Africa.	376-377
What are the prospects for South Africa in the immediate future?	Uncertain. As of the turn of the century, apartheid was dismantled and the new black-majority government was continuing to build the country's political future; but dissension within the white community and conflict from many black sectors were continuing problems facing the post-Mandela government elected in 1999.	377

South African Economic Geography

| How did the discovery of diamonds open up the South African interior? | The finds around Kimberley in the 1860s attracted thousands of fortune hunters and laborers as well as railroad construction and a thriving mining/manufacturing industry. | 377-378 |
| Besides diamonds and gold, name the other assets that supported South Africa's economic development over the past century. | Farming flourished in the favorable natural environment; cheap labor was plentiful; several other rich mineral deposits were exploited, including iron, copper, nickel, tin, and chromium; a major metals fabrication industry arose; a fine transport network matured; foreign investment flowed in and white immigration continued (at least through the 1970s); thriving cities mushroomed. | 378-379 |

| Was the country developed equitably or did regional economic differences arise? | Internal differences were always notable and have intensified in recent decades. The cities, booming industries, and modern farms resemble a developed country; the remainder of South Africa, overwhelmingly nonwhite, exhibits the miserable conditions typical of the rest of Subsaharan Africa. | 379 |

MAP EXERCISES

Map Comparison

1. Many of Africa's countries contain diverse cultural groups that were largely thrown together by the Europeans in the political territories that were carved out during the colonial era. Compare the maps on pp. 351 and 353, and record your observations about this fragmented pattern, its evolution over the past century, the countries that are most strongly affected, and how recent population movements in the post-independence era are adjusting some established cultural-geographic patterns.

2. Compare and contrast the maps on p. 348 with the map in Figure A in Appendix A. Why did the trans-Atlantic slave trade become such a lucrative operation, achieving mass proportions in such a short time period? Why did West Africa spawn so prominent a slave trade? How does the present social geography of the Caribbean and northeastern South America reflect the aftermath of the Atlantic slave trade (refer back to appropriate sections of Chapters 4 and 5)?

3. Carefully study the maps of South Africa on pp. 376-377, and discuss how the pattern of geographic provinces has changed since the political changes of 1994. Make observations about old patterns of ethnic/racial separation shown in Fig. 7-17, and how things have shifted in Fig. 7-18 since boundaries were redrawn for the 1994 election.

Map Construction   (Use outline maps at the end of this chapter)

1. In order to familiarize yourself with African physical geography, place the following on the first outline map:

    a. *Rivers*: Niger, Senegal, Volta, Benue, The Congo (former Zäire), Lualaba, Ubangi, Kasai, Blue Nile, White Nile, Zambezi, Cubango, Limpopo, Vaal, Orange

    b. *Water bodies*: Gulf of Guinea, Moçambique Channel, Lake Chad, Lake Volta, Lake Victoria, Lake Turkana, Lake Tanganyika, Lake Albert, Lake Malawi, Lake Kariba

    c. *Land areas:* Madagascar, Futa Jallon Highlands, Ethiopian Highlands, Drakensberg, the Sahara, the Sahel, Kalahari Desert, Namib Desert, Adamawa Massif, Bihe Plateau

2. On the second map, political information should be entered as follows:

   a. Label each country with its name.

   b. Locate and label each capital city with the symbol *.

3. On the third outline map, urban-economic information should be entered as follows:

   a. *Cities* (locate and label with the symbol ●): Dakar, Bamako, Niamey, Conakry, Freetown, Monrovia, Ouagadougou, Abidjan, Kumasi, Accra, Lomé, Porto Novo, Kano, Ibadan, Lagos, Enugu, Port Harcourt, Maiduguri, N'Djamena, Addis Ababa, Nairobi, Mombasa, Kisumu, Kampala, Jinja, Dar-es-Salaam, Mwanza, Yaoundé, Douala, Matadi, Kinshasa, Bangui, Brazzaville, Kananga, Lubumbashi, Luanda, Lobito, Lusaka, Ndola, Blantyre, Harare, Bulawayo, Beira, Maputo, Antananarivo, Johannesburg, Pretoria, Cape Town, Bloemfontein, Durban, Port Elizabeth, East London

   b. *Economic regions* (identify with circled letter):

      A - Copperbelt
      B - Great Dyke
      C - Witwatersrand
      D - Niger Delta
      E - Great Lakes region
      F - Tan-Zam corridor
      G - Transvaal
      H - Highveld

PRACTICE EXAMINATION

Short-Answer Questions

Multiple-Choice

1. The edge of the African plateau, especially in Southern Africa, is known as the:

   a) Rift Valley   b) Gondwana Mountains   c) Great Dyke
   d) Separate Development e) Great Escarpment

2. The West African country that shifted its capital from Lagos to the new centrally-located city of Abuja is:

   a) Nigeria   b) Tanzania   c) Zaïre
   d) Burkina Faso   e) Ivory Coast

3. South Africa's South Asian population is known as the:

a) Cape Coloured   b) Durbans   c) Afrikaners
d) Zulu   e) Indians

4. Nelson Mandela was the recent leader of:

a) Chad   b) Equatorial Africa   c) Tanzania
d) the Sahel   e) South Africa

5. Which of the following countries borders Lake Victoria?

a) Uganda   b) Nigeria   c) Zambia
d) Sudan   e) Somalia

6. The first Europeans to colonize what is now South Africa come from:

a) Britain   b) France   c) Belgium
d) the Netherlands   e) Germany

True-False

1. Before independence, the modern state of Zaïre was a colony of Britain.

2. Northern Nigeria houses the country's largest population cluster, the Hausa-Fulani.

3. The Copperbelt is partially located in the country formerly known as the Belgian Congo.

4. Africa is known as the "plateau continent."

5. East Africa lies at a lower general elevation than Equatorial Africa.

6. Johannesburg is South Africa's largest seaport.

Fill-Ins

1. The Ibo, Hausa-Fulani, and Yoruba peoples inhabit the country called _____.

2. "Homelands" were the cornerstones of the white-minority government policy known as _____, the successor of *apartheid* before both systems were dismantled in the mid-1990s.

3. The worst and most widespread of Africa's pandemic diseases is _____.

4. South Africa's largest city and co-capital is _____.

130

5. The country formerly called Zäire is now known as _____.

6. The large island lying to the east of Southern Africa is _____.

Matching Question on African Countries

____1.	Hausa-Fulani	A. Tanzania
____2.	Former Portuguese colony	B. The Congo
____3.	Former French Colony	C. Equatorial Guinea
____4.	One of world's fastest-growing	D. Kenya
	populations during the 1980s	E. Zimbabwe
____5.	Boers	F. Uganda
____6.	Contains the Copperbelt	G. Moçambique
____7.	Dar-es-Salaam	H. Nigeria
____8.	Great Dyke	I. Senegal
____9.	Island of Bioko	J. South Africa
___10.	Idi Amin	K. Zambia
___11.	Country named after a desert	L. Namibia
___12.	Former colony of Belgium	

Essay Questions

1. Discuss at some length the concerns of contemporary medical geography, particularly as they apply to Subsaharan Africa today.

2. The continental drift hypothesis contributes a great deal to the explanation of African hydrography (surface water patterns) and physiography. Show how one can use this approach to understand Africa's dominantly plateau-like terrain, the absence of mountain chains, the rift valleys, and the unusual courses of the realm's major rivers.

3. Perhaps no region has been more thoroughly disrupted and transformed by the colonial experience than West Africa. Discuss the major European penetrations here, the kinds of political territories that emerged, and how some of the region's disparate populations came into conflict during Nigeria's brutal civil war in the 1960s.

4. The possibility of a "Green Revolution" for Africa is widely debated today. Discuss the present condition of Subsaharan Africa's agriculture, the nature of the Green Revolution itself, and chances for success if it were widely implemented in this realm.

5. Past racial-separation policies in South Africa have been shown to have divided its population into a far more complex residential pattern than mere black/white segregation. Review some of these residential patterns, and also how electoral boundaries were redrawn along ethnic lines for the 1994 presidential election, and speculate on the continued evolution of social geography in post-apartheid South Africa.

TERM PAPER POINTERS

The "Term Paper Pointers" section of the Introduction chapter in this **Study Guide** offered suggestions about approaching research and writing on geographic realms and their components, and you may wish to consult this material if you are undertaking a report on a Subsaharan African region.

You should also be aware of the textbook's Web Site, which includes a number of direct links to other Web Sites that may be quite helpful in your research. It is: (http://www.wiley.com/college/regions2000),

The enormous variations that occur across African space make a regional geography almost indispensable: besides the standard surveys listed in the **References and Further Readings** section (Aryeetey-Attoh, Best & de Blij, Grove, Senior & Okunrotifa, and Stock), there are also systematic overviews that are quite useful (Adams et al., Ayittey, Bell, Benneh et al., Binns [both titles], Bohannon & Curtin, Chapman & Baker, Downs & Ryena, Fardon & Furniss, Gooneratne & Obudho, Khapoya, Knight & Newman, Lewis & Berry, Mehretu, Mountjoy & Hilling, Newman, O'Connor, Pritchard, Rakodi, Siddle & Swindell, and Turner et al.). Individual regions are also covered in Gooneratne & Obhudo, Harrison Church, Lemon, Marx, Mortimore, Saff, and Waldmeir. The *Economist* article ("Africa for Africans," [1996]) offers a survey of contemporary African trends as "Africa" [*Time*, 1996], Griffiths, Lamb, Maier, Middleton, and Reader. The history of Africa is treated by Christopher, Bohannon & Curtin, Curtin (both titles), Freeman-Grenville, Jalloh & Maizlish, Khapoya, Middleton, Murdock, Newman, Olson, Pakenham, Reader, and Wesseling.. Population issues, including the refugee problem, are surveyed by Benneh et al., Binns (1995), Black & Robinson, Goliber, Hodd, Huke, Jalloh & Maizlish, Marx, Murdock, Newman, Olson, Rakodi, Stren & White, and Turner et al. Urbanization is treated by Benneh et al., Rakodi, Saff, Stren & White, and Taylor. Development and modernization in Subsaharan Africa is dealt with by Bell, Benneh et al., Downs & Reyna, Fardon & Furniss, Gooneratne & Obhudo, Hodd, Martin & de Blij, Mehretu, Mountjoy & Hilling, O'Connor, Siddle & Swindell, and Taylor. Economic geography is covered in "Africa for Africans," Benneh et al., Hodd, Martin & de Blij, and Mehretu. The politico-geographical aspects of the post-independence era are covered in Ayittey, Black & Robinson, Davidson, de Blij, Fardon & Furniss, Griffiths , Marx, Pakenham, and Wesseling. The food crisis and related environmental problems are treated in Binns (1995), Foster, Harrison Church, Lewis & Berry, Mortimore, and Pritchard. Medical geography topics are treated in Curtis & Taket, Foster, Gesler, Gould, Kearns & Gesler, Learmonth, Meade, and Moon & Jones.

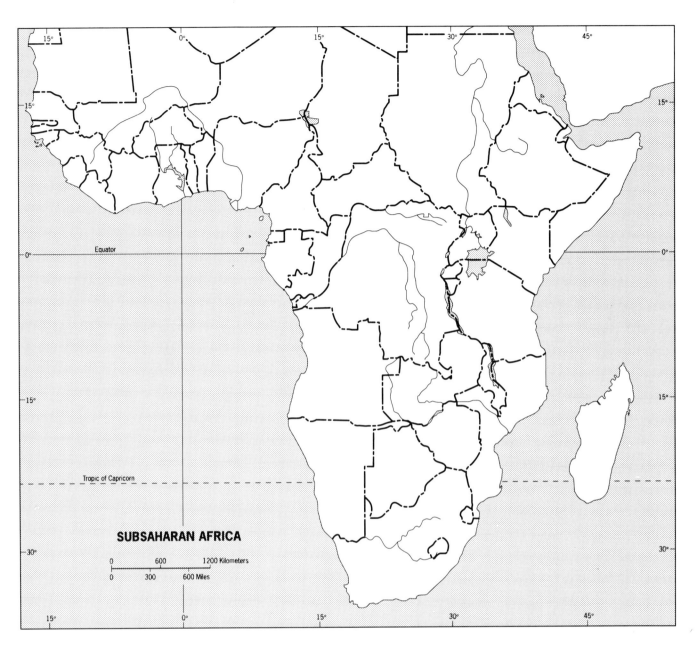

SUBSAHARAN AFRICA

0   600   1200 Kilometers

0   300   600 Miles

Equator

Tropic of Capricorn

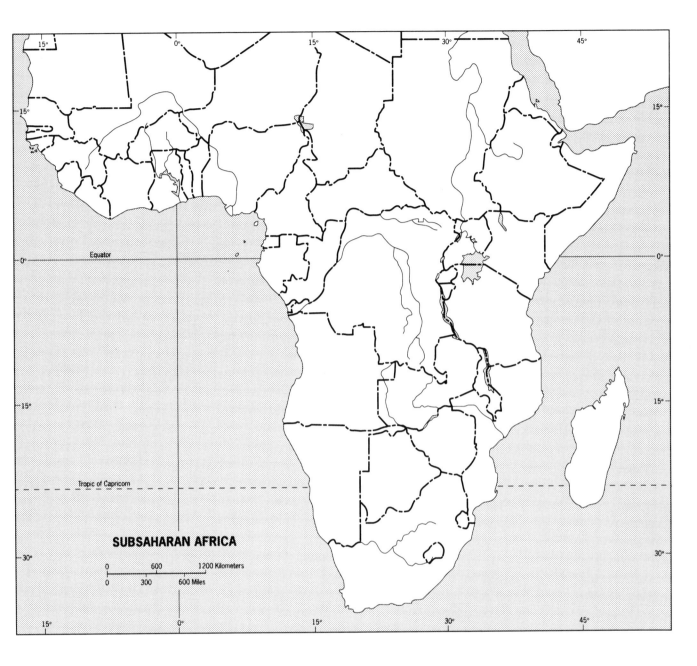

SUBSAHARAN AFRICA

Equator

Tropic of Capricorn

0   600   1200 Kilometers
0   300   600 Miles

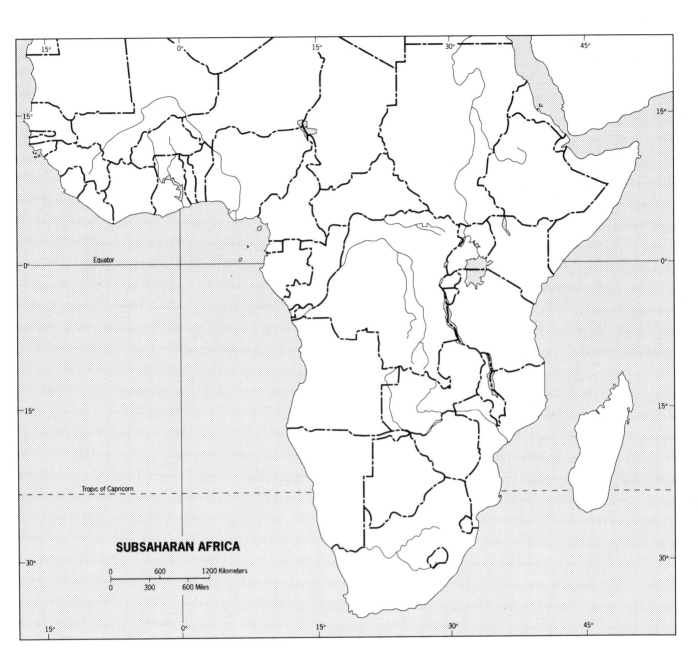

SUBSAHARAN AFRICA

Equator

Tropic of Capricorn

| 0 | 600 | 1200 Kilometers |
| 0 | 300 | 600 Miles |

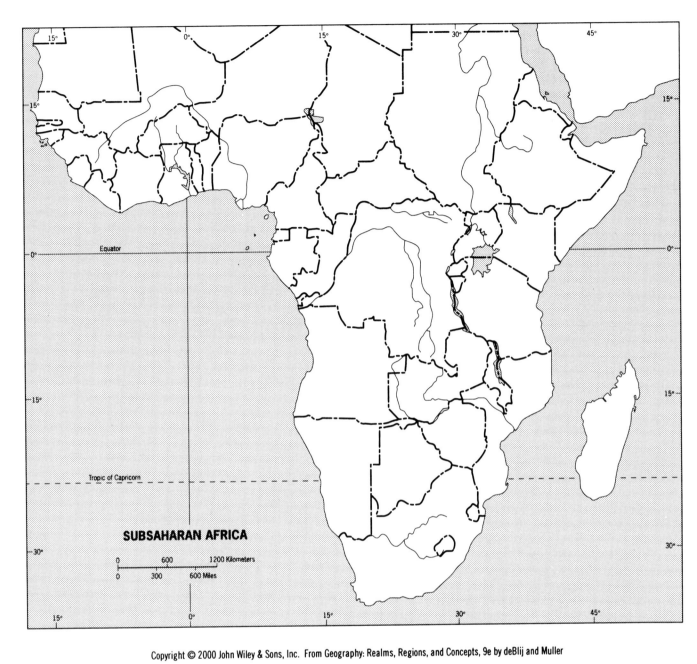

SUBSAHARAN AFRICA

Equator

Tropic of Capricorn

| 0 | 600 | 1200 Kilometers |
| 0 | 300 | 600 Miles |

CHAPTER 8

SOUTH ASIA

OBJECTIVES OF THIS CHAPTER

South Asia is the world's second largest population agglomeration, with more than 1.3 billion people residing in 2000 in India, Bangladesh, Pakistan, Sri Lanka, Nepal, Bhutan, and the Maldives. This realm's population explosion continues, and South Asia is on track to soon become the world's most populated realm and India (which surpassed 1 billion in 1999) the most populated country. Following the introduction, the rich historical geography of South Asia is traced, highlighting its significant role as a hearth of civilization and its transforming colonial experience under the British. The regional survey of South Asia begins with Pakistan and the new challenges it faces. India, the centerpiece of Chapter 8, is next, and its political, cultural, and economic geographies are reviewed. India's immense demographic problems are separately highlighted. The chapter closes with surveys of Bangladesh, Nepal and Bhutan, Sri Lanka, and the Maldives, treating geographic dimensions as well as current problems faced by each country.

Having learned the regional geography of South Asia, you should be able to:

1. Describe the physiography and the monsoon dynamics of the Indian subcontinent.

2. Trace the major stages of the realm's historical geography, particularly the dramatic transition from colonialism to independence.

3. Understand the opportunities for progress that exist in Pakistan, but are tempered by internal and external political problems that could explode at any time.

4. Understand the unique federal political structure of India and its underlying cultural mosaic.

5. Appreciate the crisis of the realm's population numbers (particularly India's).

6. Trace the recent geography of India's economic development as it has affected agriculture, manufacturing, and urbanization.

7. Understand the particular miseries of life throughout Bangladesh, and the limited prospects this country faces in the foreseeable future.

8. Understand the forces that helped Sri Lanka to make some progress during the 1970s—gains since overshadowed by the struggle for national survival as the dangerous conflict continues between the majority Sinhalese and the minority Tamils.

9. Locate the major physical, cultural, and economic-spatial features of the realm on an outline map.

GLOSSARY

Population Geography *(383-384)*

The field of geography that focuses on the spatial aspects of demography and the influences of demographic change on particular places.

Population distribution *(384)*

The way people, collectively and individually, have spatially arranged themselves in their overall environment.

Density *(384)*

The frequency of occurrence of a phenomenon within a given area.

Arithmetic density *(384)*

The number of people per unit area.

Physiologic density *(384)*

The number of people per unit area of cultivable or arable land.

Natural increase rate *(384)*

Population growth measured as the excess of live births over deaths per 1000 individuals per year. Natural increase of a population does not reflect either emigrant or immigrant movements.

Doubling time *(384)*

Period of time required for a population to double its size.

Wet monsoon *(385-box)*

The rainy season produced by an onshore airflow that dominates for weeks, occurring in the hot summer months when low atmospheric pressure over the land sucks in moisture-laden air from the adjacent cooler ocean; especially pronounced in India and Bangladesh.

Social stratification *(387)*

In a layered or stratified society, the population is divided into a hierarchy of social classes. In an industrialized society, the working class is at the lower end; elites that possess capital and control the means of production are at the upper level. In the traditional caste system of Hindu India, the "untouchables" (also known as *dalits)* form the lowest class or caste, whereas the still-wealthy remnants of the princely class are at the top.

Caste system (387;401)

The strict segregation of people according to social class in Hindu society, largely on the basis of ancestry and occupational status.

Partition (390)

The division of British India, upon independence in 1947, into India and the two (i.e., West and East) Pakistans along religious lines—with Hindus dominant in the former and Muslims in the latter two.

Forward capital (392)

Capital city located in contested territory near a border, confirming a state's desire to maintain its presence in that area. Pakistan's Islamabad is a classic example.

Centrifugal forces (401)

A term employed to designate forces that tend to divide a country—such as internal religious, linguistic, ethnic, or ideological differences.

Centripetal forces (402)

Forces that unite and bond a country together—such as a strong national culture, shared ideological objectives, and a common faith.

Irredentism (395)

A policy of cultural extension and potential political expansion by a state aimed at a community of its nationals living in a neighboring state.

Demographic transition (404-405)

Four-stage model of the changes in population growth exhibited by countries undergoing modern industrialization. High birth and death rates (Stage 1) are followed by plunging death rates (Stage 2) and then a delayed decline in birth rates (Stage 3)—producing a huge net population gain. This is followed by the convergence of birth and death rates (Stage 4) at a low overall level. See Fig. 8-10 on p. 404.

Population explosion (405)

The rapid growth of the world's human population during the past century, attended by ever-shorter doubling times and accelerating rates of increase.

Intervening opportunity (415)

In trade or migration flows, the presence of a nearer opportunity that greatly diminishes the attractiveness of sites farther away.

Natural hazard (416)

A natural event that endangers human life and/or the contents of a cultural landscape.

Tropical cyclone (416)

A hurricane (known in South Asia as a cyclone); a spiraling tropical storm that exhibits sustained wind speeds in excess of 74 m.p.h.

Insurgent state (421; see ch. 5 pp. 267-268)

Territorial embodiment of a successful guerrilla movement. The establishment by anti-government revolutionaries of a territorial base in which they exercise full control; therefore, a state-within-a-state.

SELF-TESTING QUESTIONS

Cover the right side of the page with a sheet of paper. Uncover each line after you have attempted to answer the question in the left column. If necessary, refer to textbook page(s) listed at the right.

Question	Answer	Page
Population Geography		
What does population geography study?	The distribution, growth, composition, and movement of human populations. It offers a spatial view of demography.	383-384
Define *density, arithmetic density,* and *physiologic density.*	Density is the frequency of occurrence of a phenomenon in a region; arithmetic density is the number of people per unit area; physiologic density is the num-ber of people per unit of arable land.	384
South Asian Characteristics		
Which countries constitute this realm?	India, Pakistan, Bangladesh, Nepal, Bhutan, Sri Lanka, and the Maldives.	382

Describe South Asia's three major physiographic regions.	The northern mountains, the central riverine lowlands, and the southern peninsular plateaus.	384
What is the North Indian Plain?	The belt of alluvial lowlands stretching between the Indus Plain on the west to the Brahmaputra Valley on the east.	384
What is the Punjab?	The low hills between the North Indian and Indus plains, the so-called "land of five rivers" astride the India-Pakistan border.	384
What is the course of the Indus River?	It rises in Tibet, crosses Kashmir, bends southward to receive its major tributaries from the Punjab, and then flows south-west through its lower course to reach the Arabian Sea near Karachi.	384
What are the Eastern and Western Ghats?	The hills that descend from the interior Deccan plateau to the narrow coastal plains of peninsular, southern India. The Western Ghats are well watered by the wet monsoon, and the Malabar/Konkan coastal strip at its foot is one of India's most agriculturally productive and populous areas.	384; 385-box

Historical Geography

Who were the Aryans? What was their impact on the emergence of India's modern culture?	Invaders from western Asia who conquered the early Indus Valley civilization beginning around 3500 B.C., but who adopted many of its innovations and pushed settlement frontiers east into the Gangetic Plain and south into the center of the peninsula. India's culture developed from this beginning, including the Hindu religion and its rigid social stratiification scheme–the *caste system*.	386-387
What lasting impact did Buddhism make?	It was dominant during the Mauryan Empire (3rd century B.C.-2nd century A.D.) but soon declined in South Asia, only remaining strong in Sri Lanka (formerly Ceylon) where it still prevails; Buddhism today is mainly centered in East and Southeast Asia (see Fig. 6-2).	387-388
What was the lasting impact of Islam?	After the 10th century it was a strong influence, driving out Buddhism but not Hinduism, which remained dominant in India's Ganges core area; Muslims remain a sizeable minority (more than 12 percent) in India—and form overwhelming majorities in Pakistan and Bangladesh.	388

Describe the realm's colonial-era experience.	The British quickly emerged as the dominant colonial power, first through the East India Company and then outright political control from 1857 to 1947. The British introduced many innovations, but forced the colonial economy to become a raw-material producer subservient to the English master.	388-389
Describe the events surrounding the 1947 partition of British India.	The British left India in 1947; however, before withdrawing they separated their former territory into Hindu-dominated India and the two Islamic Pakistans, thereby inducing mass migrations, conflict, and social stresses that still exist.	390

Pakistan

What factors unite Pakistan as a state?	Leaders turned to Islam to bind the country. Pakistan is one of the world's most theocratic states.	392
Where is the country's Muslim stronghold?	In the Pakistani portion of the Punjab, especially around Lahore near the Indian border.	392-393
What are some examples of irredentism in Pakistan?	The Pathans along the Afghanistan border in the northwest; the Kashmir dispute further compounds the country's centrifugal forces.	394-box; 395-396

India's Political Geography

Describe India's federal political structure.	This most populous of the world's federal states has 25 internal States, 6 Union Territories, and 1 National Capital Territory, with boundaries largely drawn to respect linguistic differences.	397
What is India's present linguistic status?	Hindi is the most widely spoken of the fourteen "official" languages, used by over one-third of the people (there are hundreds of local languages); English has emerged as the *lingua franca*, the chief language of the business world. See Fig. 8-5.	399
What are *centrifugal* and *centripetal* forces? How do they operate in India today?	Disunifying and unifying forces, respectively. A persistent centrifugal force is the stratification of society into castes. Centripetal forces include Hinduism, democracy, good communications, and flexibility to accommodate the country's many regional groups (at least through the late 1990s).	401-403

India's Demographic Trends

What is the current *doubling time* for India's population?	39 years; in the late 1990s, the annual growth rate was 1.8%. India is second only to China in total population.	404
Which country of South Asia is growing most slowly?	Sri Lanka has the lowest population growth, and also the highest per-capita GNP.	404
Is the Demographic Transition Model applicable to India?	Possibly. However, India's population is already much too large, and the pivotal area of the Ganges Lowland remains extremely overpopulated.	405-406
Has the rate of urbanization been as rapid in India as elsewhere in the less advantaged realms?	Surprisingly not, though the pace is quickening today. More than 260 million (26 percent of the population) now reside in urban areas, most in overcrowded communities that can hardly be expected to absorb the additional millions who arrive annually in the cities—a crisis that has just begun to be dealt with.	406

India's Economic Development

Why is Indian agriculture so stagnant?	Because inefficient traditional farming methods are widespread, transportation is poor, and because land continues to be divided into very small plots.	410-412
Has the Green Revolution improved farm productivity?	Yes, but the population explosion since the 1960s has prevented the closing of the gap between population and food supply; it remains a precarious situation.	412
How has industrialization progressed?	Slowly, despite good coal and iron ore deposits; investments continue but lagging development patterns are very difficult to break.	412-414

Bangladesh

Describe the country's agricultural situation.	Fertile alluvial soils permit highly intensive farming, tea and jute for cash, rice and wheat for subsistence.	416
What are some problems with Bangladesh's infrastructure?	The national communications system that was destroyed in the 1971 revolution has never been repaired; the road and rail system is inadequate; economic development is stagnating.	417

Sri Lanka

Which two groups are in conflict here?	The Buddhist, Aryan, *Sinhalese* majority vs. the Hindu, Dravidian, *Tamil* minority.	419
What is the situation in the Jaffna Peninsula?	The Tamil *insurgent state* continues to be an area of conflict, even after Sri Lankan government forces drove the Tamils out of their headquarters at Jaffna in 1995.	421-422

MAP EXERCISES

Map Comparison

1. Compare the physiographic maps of South Asia and Africa (pp. 380 and 334), noting key similarities and differences. Describe India's position within the continental drift scheme (Fig. 7-3, p. 339) and its present situation within the tectonic-plate framework (Fig. I-4, p. 9).

2. The linguistic map of India is a result of its cultural evolution over the past 3000 years. Analyze that map (Fig. 8-5, p. 387), accounting for its major differences within the broad framework of the historical-geographic forces that shaped the subcontinent's division into Indo-Aryan and Dravidian cultural spheres.

3. Environmental conditions are frequently associated with crop concentrations in subsistence economies. Compare the map of India's agriculture (p. 411) with the distributions of its rainfall regime (p. 12) and climates (p. 14-15), and make observations about the relative productivity of India's regions and the associations of the following crops with specific climate types: cotton, rice, and wheat. Do your conclusions also hold for Sri Lanka (map p. 420)?

4. To get an idea of the 1931-1991 shifts of India's Muslim population, compare the three maps of Fig. 8-6 (p. 391), and describe the overall trends in a spatial context. Which areas now hold a sizeable Muslim population that did not in the early twentieth century?

Map Construction *(Use outline maps at the end of this chapter)*

1. In order to familiarize yourself with South Asian physical geography, place the following on the first outline map:

    a. *Rivers*: Indus, Ganges, Brahmaputra, Hooghly, Meghna, Godavari, Tapti, Narmada, Mahanadi, Yamuna, Sutlej

b.   *Water bodies*: Bay of Bengal, Arabian Sea, Palk Strait, Ganges-Brahmaputra Delta, Rann of Kutch

c.   *Land bodies*: Sri Lanka, Sind, Kathiawar Peninsula, Cardamom Upland, Thar Desert, Deccan Plateau, Malabar Coast, Coromandel Coast, Konkan Coast, Golconda Coast, North Indian Plain, Punjab, Indus Plain, Baluchistan, Assam Uplands, Chota Nagpur Plateau

d.   *Mountains*: Himalayas, Hindu Kush, Karakoram Range, Western Ghats, Eastern Ghats, Vindhya Range, Khasi Hills

2. On the second map, political-cultural information should be entered as follows:

a.   Draw in the internal State boundaries of India (text p. 398).

b.   Draw in the language regions mapped on p. 387 and color them in appropriate shades.

c.   Compare these two maps and offer conclusions. As an optional additional exercise, consult Fig. 8-6 (p. 391) and map in concentrations of Muslim minorities within India; how do these clusters relate to linguistic patterns and how have they changed since 1931 (Fig. 8-5 [p. 387])?

3. On the third outline map, urban-economic information should be entered as follows:

a.   *Cities* (locate and label with the symbol ●): Mumbai (Bombay), Calcutta, Madras, New Delhi-Delhi, Bangalore, Ahmadabad, Varanasi, Hyderabad (once each for India and Pakistan), Amritsar, Kanpur, Pune, Bhilai, Bhopal, Patna, Jammu, Srinagar, Agra, Chandigarh, Nagpur, Mysore, Pondicherry, Madurai, Jamshedpur, Lucknow, Jaipur, Karachi, Lahore, Rawalpindi, Islamabad, Multan, Peshawar, Faisalabad, Dhaka, Cox's Bazar, Chittagong, Khulna, Kathmandu, Thimphu, Colombo, Jaffna, Anuradhapura, Male

b.   *Economic Regions* (identify with circled letter):

        A -   Punjab
        B -   Assam
        C -   Bihar-Bengal Industrial Region
        D -   Chota Nagpur Industrial Region
        E -   Maharashtra-Gujarat Industrial Region
        F -   Delhi conurbation
        G -   Sind
        H -   Baluchistan
        I -   Hindustan
        J -   Vale of Kashmir

PRACTICE EXAMINATION

Short-Answer Questions

Multiple-Choice

1. The majority religion practiced in Sri Lanka is:

   a) Dravidian   b) Aryan   c) Islam
   d) Buddhism   e) Hinduism

2. The partitioning of Hindu India from Muslim Pakistan occurred in:

   a) ca. 460 B.C.   b) 1857   c) 1947
   d) 1971   e) 1984

3. Which of the following is not located in Pakistan?

   a) Malabar Coast   b) Sind   c) Punjab
   d) Baluchistan   e) Islamabad

4. Which Indian city was renamed Chennai in the late 1990s?

   a) Calcutta   b) Gandhi   c)Madras
   d) Karachi   e) Bombay

5. Which country does not border India?

   a) China   b) Pakistan   c) Bangladesh
   d) Nepal   e) Afghanistan

6. The doubling time for India's population is approximately:

   a) 32 years   b) 40 years   c) 97 years
   d) 109 years   e) 119 years

True-False

1. The Pakhtun irredentist movement in northern Pakistan is based on cultural linkages to neighboring India.

2. Sikhism developed and is still based in the Punjab region.

3. Dravidians are in the majority in the population of the island-nation formerly called Ceylon.

142

4. The Maldives are located in the Bay of Bengal.

5. Karachi is located at the mouth of the Indus River.

6. The Ganges and Brahmaputra rivers join together to form a double delta in Pakistan.

Fill-Ins

1. The social stratification that dominates Hindu India is known as the _____ system.

2. The narrow lowland of far southeastern India that borders the Indian Ocean and contains the city of Madras, is called the _____ Coast.

3. The amount of arable land per person is known as the _____ density.

4. The dominant ethnic group of Sri Lanka is called the _____.

5. The body of water to the west of India is the _____ Sea.

6. India's first prime minister after independence was _____.

Matching Question on South Asia

____1.	Aryan Buddhist	A. Sepoy Rebellion
____2.	Bihar-Bengal Industrial Region	B. Chennai
____3.	Pakistan's northwestern frontier	C. Aśoka
____4.	Forward capital today	D. Bangladesh
____5.	India's *lingua franca*	E. Sind
____6.	Dravidian language	F. Sinhalese
____7.	India takeover by	G. Tamil
	British government	H. Islamabad
____8.	Formerly Bombay	I. English
____9.	Indus Delta, centered on Karachi	J. Calcutta
___10.	Mauryan Empire	K. Mumbai
___11.	Disastrous 1991 hurricane	L. Khyber Pass
___12.	Formerly Madras	

Essay Questions

1. The Kashmir problem continues to be an unresolved crisis of major politico-geographical significance to South Asia. Trace the origin of this conflict, its evolution through the 1980s, and the latest events that have occurred since 1990.

2. Discuss the evolution of Hinduism as the leading religion and social system in India. Trace the various influences and challenges of the Mauryan Empire, the Mogul Empire, the British colonial era, and the course of Hinduism since independence.

3. Discuss the current agricultural geography of India. Draw a sketch map that shows the regional distributions of rice and wheat, and discuss their relationship to the dynamics that shape the country's annual wet monsoon.

4. India's industrialization and urbanization have proceeded rather slowly over the past generation. Discuss the reasons for the leisurely manufacturing expansion, and why urbanization has accelerated markedly since 1980.

5. Bangladesh is undoubtedly plagued by more miseries than almost any other country in the world. Review the environmental, cultural, and economic forces that created this situation, and discuss the relationship between the country's population geography and hopes for future development and modernization.

TERM PAPER POINTERS

The "Term Paper Pointers" section of the Introduction chapter in this **Study Guide** offered suggestions about approaching research and writing on geographic realms and their components, and you may wish to consult this material if you are undertaking a report on a South Asian region. You should also be aware of the textbook's Web Site, (http://www.wiely.com/college/regions2000), which includes direct links to a number of other Web Sites which may be quite helpful in your research.

Size and diversity of the realm as a whole, and within India in particular, makes a regional geography an indispensable reference. In the **References and Further Readings** section, Farmer and Johnson (1983) are the newest sources; Spate & Learmonth is a classic work, and Spencer & Thomas and Chapman & Baker are useful general surveys. Dutt & Geib and Muthiah are the compilers of comprehensive recent atlases, and Robinson provides a broad informational overview. For India alone, the key geographical works (in addition to those just cited) are Corbridge (1997), Deshpande, Johnson (1979), Noble & Dutt, Sopher, the splendid historical atlas compiled by Schwartzberg—which also includes countries of the Indian perimeter; more general works on India and its current problems are Crossette (1993), Jaffrelot, Khilnani, Lall, Mehta, Paz, Rushdie, Tharoor, "Time to Let Go," and Wolpert. Pakistan is treated by Samad; Bangladesh by Baxter, Er Rashid, Haque, and Johnson (1982). As for the smaller countries, Sri Lanka is covered in Isaac and McGowan, and Nepal and Bhutan in Crosette (1995), Ives & Messerli, Karan, and Karan & Ishii (both titles).

As for topical treatments, cultural geography is reviewed in Corbridge & Harriss, Dumont, Hasan, Lipner, Lukacs, Noble & Dutt, Samad, and Sopher. The more specialized topic of religious geography is covered in Corbridge & Harriss, Hasan, Jaffrelot, Lipner, and Sopher. Aspects of political geography are examined by Burns (1998), Corbridge & Harriss, Mitra, Samad, van der Veer, and Wiring (both sources). Urban geography is treated in Costa et al. and Dutt. Environmental topics are treated in Haque, Ives & Messerli, and Thapar. Population geography is treated in Dumont, Kosinski & Elahi, Lukacs, McFalls, Newman, Sanderson & Tan, van der Veer, and Zachariah & Rajan. Recent conflicts and upheavals are covered in Corbridge & Harriss, Isaac, Jaffrelot, McGowan, Mitra, Samad, van der Veer, and Wiring (both titles).

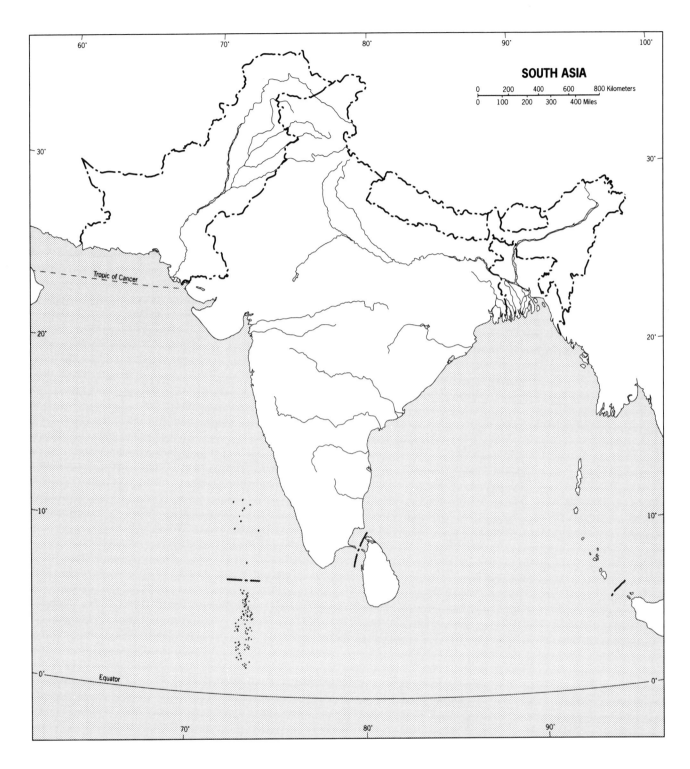

SOUTH ASIA

0   200   400   600   800 Kilometers
0   100   200   300   400 Miles

30°   30°

Tropic of Cancer

20°   20°

10°   10°

0°   Equator   0°

60°   70°   80°   90°   100°

70°   80°   90°

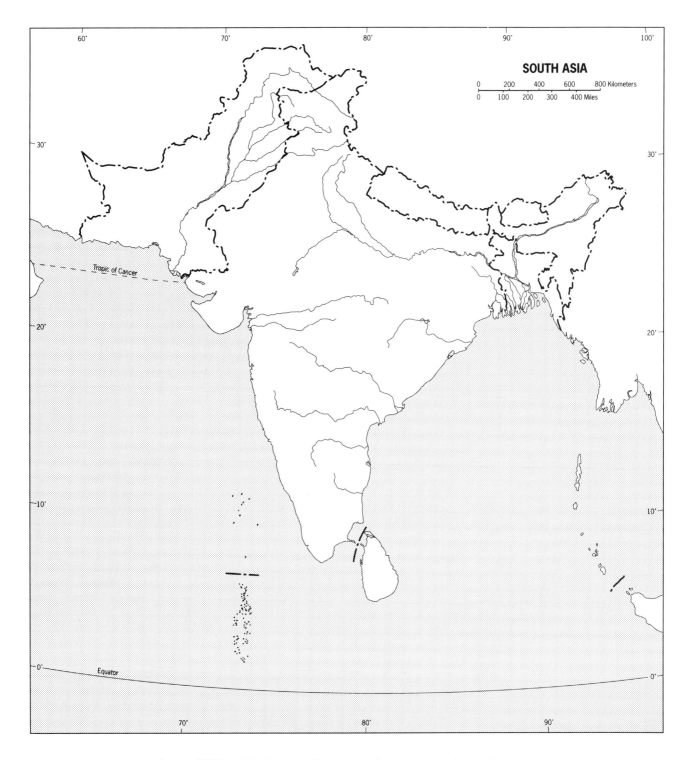

0   200   400   600   800 Kilometers

0   100   200   300   400 Miles

Tropic of Cancer

Equator

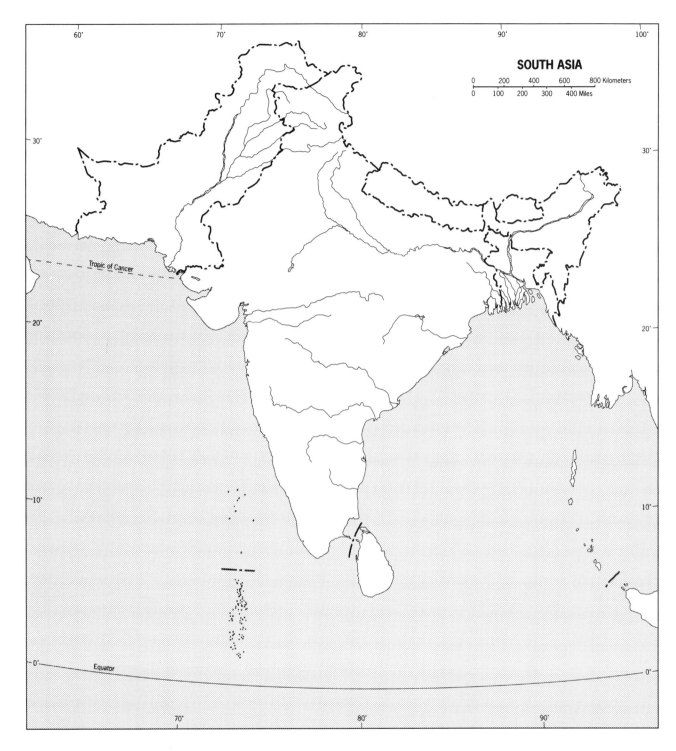

SOUTH ASIA

0   200   400   600   800 Kilometers
0   100   200   300   400 Miles

Tropic of Cancer

Equator

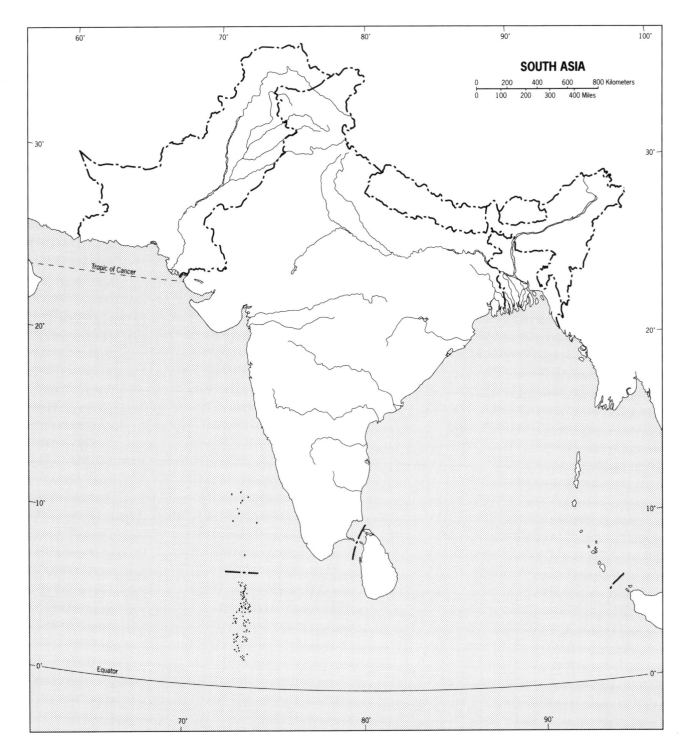

SOUTH ASIA

0   200   400   600   800 Kilometers

0   100   200   300   400 Miles

30°

Tropic of Cancer

20°

10°

0°   Equator

60°   70°   80°   90°   100°

CHAPTER 9

EAST ASIA

OBJECTIVES OF THIS CHAPTER

Chapter 9 covers East Asia, focusing largely on China—the now-awakened giant that today arouses a great deal of interest and concern. Not only is China the world's largest country in population size and capable of becoming an international superpower in the foreseeable future; it is also in the throes of upheaval. The introduction discusses China's current dimensions, its role as the last surviving empire, and the contradictions that mark its emergence on the global scene. China's natural environment is then discussed. Historical geography follows, tracing the critical events of dynastic China, the convulsions of the colonial era, and the communist transformation of the past five decades. Next comes an extended survey of China's regional human geography, highlighting the contents of the major subdivisions of the country, with emphasis on the internal structure and functioning of the spatial components of China Proper. We highlight China's developmental experiences since the 1949 Communist Revolution, the modernization program pursued by the post-Mao leadership since the late 1970s, and China's prospects for the immediate future. Included is a look at China's booming Special Economic Zones, and the former British Crown Colony of Hong Kong, reunited with China (and renamed Xianggang) in 1997. We then focus on the Jakota Triangle—consisting of Japan, South Korea, and Taiwan. Japan's rise to a position of global prominence is traced. Working our way along the western Pacific Rim, we next come to South Korea, one of the realm's *economic tigers*. Taiwan, another tiger, follows—its political future remaining to be settled with China (which claims the island-nation).

Having learned the regional geography of East Asia, you should be able to:

1. Understand the basics of the geography of the realm's development and modernization.

2. Appreciate the growing prominence and changing role of China in global affairs.

3. Appreciate the long tradition of Chinese culture and the tremendous upheavals that its adherents have experienced during the past century and a half.

4. Describe the contents and spatial organization of each of China's major regions, particularly the components of China Proper.

5. Understand the background and goals of China's ongoing modernization program—as well as its important spatial expressions—so that you can monitor and evaluate its progress in the future.

6. Comprehend the challenges now facing China as it struggles to reconcile its economic and political directions.

7. Understand the western Pacific Rim's place in today's world as a newly emerging assemblage of regions, composed of discontinuous yet interconnected economic giants and fast-rising economic tigers.

8. Trace Japan's modernization trend and its impact from the Meiji Restoration (1868) to the present.

9. Describe Japan's resource base and the broad regional structuring of its productive activities.

10. Describe the basic geographic contents and spatial organization of South Korea and Taiwan.

11. Locate the leading physical, cultural, and economic-spatial features of the realm on an outline map.

GLOSSARY

Pacific Rim *(426-427)*

A far-flung group of countries and parts of countries (extending clockwise on the map from New Zealand to Chile) sharing the following criteria: they face the Pacific Ocean; they evince relatively high levels of economic development, industrialization, and urbanization; their imports and exports mainly move across Pacific waters.

Pinyin system *(427-box)*

The standard form of Chinese established by the government in 1958 that is based on the pronunciation of the language in the north of China.

"People of Han" *(432-table;438-439)*

Because the Han Dynasty (207 B.C.-A.D. 220) was the formative period of China's traditional culture, most ethnic Chinese still refer to themselves in this way.

Confucianism [Kongfuzi] *(438-box)*

Ideals deeply ingrained in the Chinese culture and national character, which still linger after more than a half century of communism.

Sinicization *(439)*

Giving a Chinese cultural imprint; Chinese acculturation.

Extraterritoriality *(442-box)*

Legal concept that the property of one state lying within the boundaries of another actually forms an extension of the first. Used in colonial-era China by European diplomats and others to carve out neighborhood enclaves that were off-limits to residents of the host country.

Manchuria *(442-box)*

The foreign term for *Northeast China* (Liaoning, Jilin, and Heilongjiang Provinces), which the Chinese do not recognize.

Manchukuo *(442-box)*

The Japanese name for Manchuria during their Empire days of the 1930s and early 1940s.

The Long March *(443)*

The year-long exodus of the communists, who were driven out of China Proper in 1934, through the difficult Chinese interior; from their new interior mountain refuge they eventually gathered sufficient strength to emerge again in the late 1940s and finally defeat the Nationalists in 1949.

Core area *(452)*

In geography, a term with several connotations. Core refers to the center, heart, or focus. The core area of a nation-state is constituted by the national heartland, the largest population cluster, the most productive region, and the part of the country with the greatest centrality and accessibility—probably containing the capital city as well.

Qanats *(460;282)*

Underground tunnels used to transport water from mountains into nearby, lower-lying desert areas.

Geography of development *(463)*

The field of geography concerned with spatial aspects and regional expressions of development.

Development *(463)*

The economic, social, and institutional growth of national states.

"Great Proletarian Cultural Revolution" (463-464)

The reign of terror in the name of revolutionary zeal unleashed by Mao Zedong in the middle and late 1960s, wherein the excesses of the militant Red Guards intimidated the population to purge China of noncommunist traditions (and conceal the disastrous economic setbacks of the "Great Leap Forward").

Special Economic Zone [SEZ] (464-465)

Chinese coastal manufacturing and export center, established in the 1980s and 90s, to attract foreign investments and technology transfers through special tax benefits and additional economic incentives.

Economic tiger (466)

One of the burgeoning beehive countries of the western Pacific Rim. Following Japan's route since 1945, these countries have experienced significant modernization, industrialization, and Western-style economic growth since 1980. Three leading economic tigers today are South Korea, Taiwan, and Singapore. Term is increasingly used more generally to describe any fast-developing economy (e.g., Ireland's growing reputation as the "Celtic Tiger").

Regional state (469-box)

A "natural economic zone" that defies political boundaries, and is shaped by the global economy of which it is a part; its leaders deal directly with foreign partners and negotiate the best terms they can with the national governments under which they operate.

Buffer state (472)

Remnant of bygone era in which European colonial powers carved out spheres of influence, particularly in Asia. To avoid confrontation, certain territory was left relatively empty between such spheres as a cushion or "buffer." Mongolia evolved as a buffer state between China and Russia. Today it remains landlocked, isolated, and vulnerable, as do such other Asian buffer states as Afghanistan and Tibet.

Jakota Triangle (473)

Japan, South Korea, and Taiwan form this East Asian region of countries in the western Pacific Rim, which exhibits great cities, huge exports, global linkages, and rapid development.

Archipelago (474)

A chain of islands grouped closely together—as in the case of Japan.

Relative location (479)

The regional position or situation of a place relative to the position of other places. Distance, accessibility, and connectivity affect relative location.

Areal functional organization (479)

Philbrick's model of regional organization based on 5 interrelated ideas that activities concentrate in a locality, interact with other places, and participate in the evolution of a multi-level hierarchy of areal organization.

Physiologic density (483; 384)

The number of persons per unit of arable or cultivable land; Japan exhibits one of the world's highest physiologic densities.

Aquaculture (483-484)

The use of a pond or any other artificial body of water to grow food products, including fish, shellfish, and even seaweed; method often also used in shallow bays and estuaries.

Regional complementarity (487)

Exists when two regions, through an exchange of raw materials and/or finished products, can specifically satisfy each other's demands.

State capitalism (488)

Government-controlled corporations competing under free-market conditions, usually in a tightly regimented society (such as South Korea).

SELF-TESTING QUESTIONS

Cover the right side of the page with a sheet of paper. Uncover each line after you have attempted to answer the question in the left column. If necessary, refer to textbook page(s) listed at the right.

Question	Answer	Page
The Pacific Rim		
What are the major similarities of the western Pacific Rim countries?	First, they all adjoin the Pacific Ocean. Second, they are comparatively highly urbanized, which reflects relatively high levels of economic development. Third, they are either island countries, peninsulas, or coastal strips. Fourth, most of their trade moves across Pacific waters.	426-427
Where have some of the most spectacular developments occurred in the western Pacific Rim?	In the countries lining the western Pacific: Japan, South Korea, Taiwan, Hong Kong (Xianggang), coastal and especially southern China.	426-427
China's Characteristics		
What is the *pinyin* system?	The standard form of Chinese, which represents a more modernized version of the language; used in this text for all Chinese names.	427-box
What does China's general climate pattern look like?	Mild *C* climates dominate in the Southeast, yielding to colder, drier climates towards the west; *D* climates prevail in the north; *H* or highland climates in the high-altitude west.	428; 437
What are China's three most important rivers?	Huang He (Yellow), Chang (Yangtze [Yangzi]), and Xi (West).	428-429
What are the main farming/population clusters of China?	The North China Plain, the lower Chang (Yangzi) Basin, the Xi Basin, and the Sichuan Basin of the interior.	429
Regions of East Asia		
What are the regions that comprise the East Asia realm?	China Proper, Xizang (Tibet), Xinjiang, Mongolia, and the Jakota Triangle (Japan, South Korea, and Taiwan).	434

Historical Geography

When and where did the earliest Chinese culture develop?	The Xia Dynasty formed around 2200 B.C., in the core area centered on the lower reaches of the Yi-Luo River Valley.	432-table; 438-439
Who was China's most influential philosopher and teacher?	Confucius (Kongfuzi or Kongzi), who lived from 551-479 B.C., who educated and urged the poor to assert themselves towards feudal lords, and taught that human virtues should determine a person's place in society.	438-box
What were the major accomplishments of the formative Han period (206 B.C.-A.D. 220), that made it the "Roman Empire" of East Asia?	Unity and stability established over a vast area (see Table 9-1); trade routes secured; control over nomads; private property rights emerged; strong military power; external commerce; advances in science and the arts.	432-table; 438-439
Why was China's self-assured isolationism shattered in the 19th century?	The European powers economically destroyed China's handicraft industries, and their political influence in parts of China bordered on colonialism, a superiority enhanced by encouraging the use of opium that weakened the fabric of Chinese society.	440
What is the concept of *extraterritoriality*?	The idea that a foreign power could have territorial enclaves in another country which were legal extensions of that foreign state. Thus foreign states were immune from the jurisdiction of the country in which they were based.	442-box
List China's political fortunes since 1900.	1900 Boxer Rebellion began the end of colonialist meddling; 1911 overthrow of monarchy led to republican period of government led by Sun Yat-sen and then Chiang Kai-shek; the Japanese dominated from the late 1930s to 1945; civil war from 1945 to 1949 led to communist takeover and Nationalist flight to Taiwan.	442-444
What was the Long March?	The long, interior exodus of the communists in 1934-1935, who managed to avoid their Nationalist antagonists and eventually emerge victorious in the late 1940s.	443

| When did communism become the controlling force of the Chinese state? | In 1949, when Mao Zedong proclaimed the success of the revolution and the birth of the People's Republic of China. | 444 |

Political Map and Population

How is China's political framework divided?	There are 22 provinces, 5 autonomous regions, 4 central-government-controlled municipalities (*shi's*), and 1 special administrative region..	445
What is China's most populous province?	Sichuan, including its outer borders. Population declined on paper for Sichuan when Chongqing, population 30 million, was recently made a separate *shi*. In actuality, however, Sichuan and its hinterland remain the country's most populous province.	445-446
How large is China's population?	1.3 billion, about one-fifth of humankind.	447
How has the attitude of the government toward population control changed since the 1970s?	In the 1970s, couples were instructed to have one child only, and penalized if they had more. Negative effects of the one child policy were soaring abortion rates (many late in pregnancies) and a rise in female infanticide. Attitudes have now relaxed slightly, and a modest rise in growth rates has resulted.	447
Which ethnic group forms the largest population clusters in China?	The Han Chinese (the ethnic Chinese) form the largest and densest clusters, as shown in Fig. 9-9, p. 448.	448
What percentage of the Chinese live in rural areas?	Approximately 70% of the population.	449

China's Regions

| What are the leading industrial resources of the Northeast? | Iron ore and coal in the Liao Basin, supporting heavy industrial production in both Anshan and Shenyang; Daqing oilfield near Harbin. | 450 |
| Why does the Northeast China Plain look like a rustbelt? | Industrial manufacturing that flourished from the 1950s to the 1970s under communist planning declined dramatically under the new, market-driven economic restructuring of the 1980s. | 450-451 |

Where is the North China Plain, and what is its significance?	The lower basin of the Huang River, extending as far north as Beijing; it is one of the world's most fertile, productive, and densely-populated farming regions. See Fig. 9-12, p. 453.	452
Why is the North China Plain an excellent example of a national core area?	It is densely populated, agriculturally productive, and contains the capital and other large cities, a leading industrial complex, and several major ports.	452
Which of China's major rivers is most navigable?	The Chang; Shanghai developed at its delta.	454
When and how was Xizang (Tibet) occupied by China?	In 1959, by military force, after years of economic interference and frontier settlement.	460
Describe the human geography of Inner Asian China.	A sparse population inhabits the dry environments of Inner Mongolia and Xinjiang; in the far west, a sizeable majority are Muslims, with cultural affinities to the republics of neighboring Turkestan.	460-462; 472

Geography of Development

How is development measured?	Indicators of development are GNP per person, occupational structure of the labor force, energy consumption, transportation and communication levels, amount of metals required each year, productivity per worker, and rates of literacy, nutrition, savings, and the like.	463
What are the stages in Rostow's development model?	(1) Traditional society; (2) Preconditions for takeoff; (3) Takeoff; (4) Drive to maturity; (5) High mass consumption. A new stage might be added: the postindustrial stage of ultra-sophisticated high technology.	463

China's Economic Zones

What are Special Economic Zones (SEZs)?	Manufacturing and export centers in coastal China that lure foreign investment capital and factories through tax benefits and other economic incentives.	464
Besides SEZs, what other economic zones were introduced in recent years?	Fourteen other cities (Open Cities), where foreign investment is encouraged.	465

Why has Shenzhen been the most successful of China's Special Economic Zones (SEZs)?	Shenzhen lies adjacent to Xianggang (Hong Kong), has access to port facilities, and a labor force willing to work for lower wages than in Hong Kong. Hong Kong, Japan, the U.S., Taiwan, and Singapore all invest in Shenzhen.	466
What are the three entities that constitute Xianggang (Hong Kong)?	The island named Hong Kong, the Kowloon Peninsula, and the New Territories—the latter two being on the mainland.	466
When did Hong Kong become the Xianggang Special Administrative Region (SAR)?	At midnight on June 30, 1997, following China's takeover after the departure of the British.	466
How did Hong Kong become an economic tiger?	Its large supply of labor and foreign investment helped to launch a textile as well as light manufacturing industries, which exported massively to foreign countries.	467

The Jakota Triangle

Which countries are included in this region?	Japan, South Korea, and Taiwan.	473
What characteristics do the countries of the Jakota Triangle share?	Burgeoning cities, high consumption of raw materials, huge exports, global linkages, and rapid development.	473

Japan's Modernization

When were the old rulers of Japan overthrown, paving the way for reform?	In 1868, during the rebellion called the Meiji Restoration—the return of the enlightened rule.	473-474
How were the Japanese able to transform their society since 1868?	By careful domestic planning rather than through foreign intrusion; building on internal resources, the Japanese expanded to create an empire.	474-478

Japan's Spatial Organization

Name the four main islands of the Japanese archipelago.	Honshu (the largest), Kyushu, Shikoku, and Hokkaido.	474

Which nation forced open the modern era of Japanese international contact and trade?	The United States in the 1850s, through a show of naval strength, extracted the necessary agreements.	478
Where did Japan's early industrial growth concentrate and why?	Largely in the coastal cities, because raw materials were transported by water and labor forces were already located there.	478
What are Japan's shortcomings in local industrial resources?	Inferior coal, and very little iron ore; oil supplies are practically nonexistent, forcing importation and heavy reliance on hydroelectric and nuclear energy production.	478
What role did relative location play in Japan's post-World War II development?	The U.S. had become the economic and political focus of the world; Japan's location on the western Pacific coast was no longer remote.	479
What percentage of Japan's land is considered habitable?	An estimated 18 percent.	479
List the main ideas of Philbrick's principle of *areal functional organization.*	Human activities concentrate in a locality, interact with other places, and participate in the evolution of a multi-level hierarchy of areal organization.	479
What are the leading features of the Kanto Plain?	Japan's dominant urban and industrial region, focused on Tokyo-Yokohama-Kawasaki; fine harbor; level land and climate suitable for farming; and central location in country.	480-481
What are the leading features of the Kansai District?	Japan's second-ranking economic region; focused on the Kobe-Osaka-Kyoto triangle; busy ports; good farming area, especially for rice.	481-482
What are the leading features of the Nobi Plain?	Third-ranking producing region; focused on Nagoya, leading textile manufacturer.	482
What are the leading features of the Kitakyushu conurbation?	Japan's fourth-largest economic region; focused on 5 cities at northern end of Kyushu Island; excellently situated on trade routes to Korean and Chinese ports.	482

Japan's Food Production

How has the Japanese agricultural labor force changed since 1920?	As the economy took off, full-time agricultural employment fell from 50% of the national total in 1920 to 4% by the turn of the 21st century.	483
Why must Japan rely so heavily on food imports?	Arable land is in short supply; however, technological research for efficient agricultural methods have maximized this country's very limited farming opportunities.	483
What new food resources are being developed?	Besides the unending search for more efficient farming methods, *aquaculture* is increasingly used to supplement the food supply.	483-484

Korea

When and how did Korea become divided into North Korea and South Korea?	In 1945, at the end of World War II, the victorious Allied powers gave Korea north of the 38th parallel to the Soviet Union; territory south of the 38th parallel was given to the United States, forming South Korea.	485
Describe South Korea's growth as an economic tiger.	U.S. and then Japanese aid after the Korean War (1950-1953) launched a development boom that by the 1980s propelled the country to industrial prowess in shipbuilding, steel, automobiles, and chemicals manufacturing.	487

Taiwan

What types of goods currently dominate Taiwan's exports?	High-technology equipment, such as personal computers, telecommunications components, and precision electronic instruments.	490
What is Taiwan's greatest problem?	Its political situation. In 1949, the Chinese Nationalists made Taiwan the Republic of China. The leaders of the People's Republic of China consider Taiwan to be part of the mainland's territory, its present status that of a wayward province. Taiwan has evolved into a democratic state recently—its uncertain future hangs in the balance.	491

MAP EXERCISES

Map Comparison

1. The historical evolution of the Chinese state followed a spatial-expansion pattern associated with territorial gains. Trace these additions to China's core area in association with the dynastic traditions and trends that prevailed during the time of territorial acquisition. What spatial differences and similarities can you detect between the Chinese growth pattern (Fig. 9-3) and the expansion of Islam (Fig. 6-5), South American cultures (Fig. 5-5), United States settlement (Fig. 3-6), and the Russian Empire (Fig. 2-3)?

2. Discuss the various centripetal and centrifugal forces that currently affect the Chinese state, basing your observations on various maps in Chapter 9—especially the map of ethnic minorities in Fig. 9-9, p. 448.

3. Compare the map of Japanese manufacturing regions and livelihoods (Fig. 9-21, p. 476), with the population map of Japan, (Fig. 9-20, p. 475). Record your observations as to the correspondence between population distribution and the location of the country's primary, secondary, and tertiary industrial regions.

4. Compare the maps of South Korea (Fig. 9-23, p. 486) and Taiwan (Fig. 9-24, p. 490). Record your observations as to the internal economic geographies of these economic tigers, noting similarities and differences between spatial patterns of agricultural land use, transportation, and urbanization.

Map Construction   (Use outline maps at the end of this chapter)

1. In order to familiarize yourself with East Asia's physical geography, place the following on the first outline map:

    a. *Rivers*: Chang (Yangzi), Huang He, Wei, Xi, Pearl, Mekong, Brahmaputra, Tarim, Liao, Songhua, Ussuri, Amur

    b. *Water bodies*: East Sea (Sea of Japan), Seto Inland Sea, Tokyo Bay, Tsugaru Strait, Strait of Shimonoseki, Korea Strait, Yellow Sea, Taiwan Strait, Pearl River Estuary, Liaodong Gulf, Korea Bay, Bo Hai Gulf, East China Sea, Formosa Strait, South China Sea, Gulf of Tonkin, Bay of Bengal

    c. *Land bodies*: Hainan Island, Korean Peninsula, Shandong Peninsula, Liaodong Peninsula, Ryukyu Islands, Kurile Islands, Tarim Basin, Gobi Desert, Ordos Desert, Loess Plateau, Qinghai-Xizang Plateau, Takla Makan Desert, Junggar Basin, Qaidam Basin, Red Basin (Chengdu Plain), Liao-Songhua Lowland, Yunnan-Guizhou Plateau, North China Plain, Honshu, Hokkaido, Shikoku, Kyushu, Taiwan

    d. *Mountains*: Pamir Knot, Hindu Kush, Karakoram Range, Himalayas, Trans-Himalayas, Tian Shan, Kunlun, Altun, Altay Mountains

2. On the second map, political-cultural information should be entered as follows:

   a. Draw in all international borders on the map and label each country.

   b. Draw in each of China's province-level boundaries and label each province and autonomous region (use Fig. 9-8 as base).

   c. Plot the concentration of China's ethnic minorities (Fig. 9-9) and learn their provincial affiliations.

3. On the third outline map, urban-economic information should be entered as follows:

   a. *Cities* (locate and label with the symbol ●): Beijing, Shanghai, Xianggang (Hong Kong), Tianjin, Tangshan, Xian, Wuhan, Guangzhou, Dalian (Lüda), Shenzhen, Shenyang, Changchun, Fushun, Anshan, Harbin, Vladivostok, Lanzhou, Baotou, Taiyuan, Qingdao, Zhengzhou, Nanjing, Hangzhou, Chengdu, Chongqing, Fuzhou, Macau, Haikou, Nanning, Lhasa, Xuzhou, Jinan, Ürümqi, Yumen, Karamay, Yichang, Shantou, Zhuhai, Xiamen, Yantai, Qinhuangdao, Lianyungang, Nantong, Wenzhou, Zhanjiang, Beihai, Ningbo, Tokyo, Yokohama, Osaka, Kobe, Kyoto, Nagoya, Kitakyushu, Sapporo, Seoul, Pusan, Kwangju, Pyongyang, Taipei, Kaohsiung

   b. *Economic Regions* (identify with circled letter):

   China:

   A - Red Basin (Chengdu Plain)
   B - North China Plain
   C - Liao-Songhua Lowland
   D - Loess Plateau
   E - Pearl River Estuary
   F - Shenzhen Special Economic Zone

   Japan:

   G - Kanto Plain
   H - Kansai District
   I - Nobi Plain
   J - Kitakyushu

PRACTICE EXAMINATION

Short-Answer Questions

Multiple-Choice

1. Which movement was launched by Mao Zedong in the 1960s to rekindle enthusiasm for the Chinese brand of communism?

    a) Four Modernizations Program   b) Great Proletarian Cultural Revolution
    c) Confucianism (Kongzi)   d) Great Leap Forward
    e) The Long March

2. Which of the following regions is often called "Manchuria" by uninformed foreigners?

    a) Northeast China   b) Taiwan   c) North China Plain
    d) Red Basin of Sichuan   e) Xizang

3. Which of the following is not located on the island of Honshu?

    a) Yokohama   b) Kitakyushu   c) Nobi Plain
    d) Kyoto   e) Nagoya

4. Which of the following is not found in the Chang/Yangzi Basin?

    a) Pudong   b) Xinjiang   c) Chongqing
    d) Wuhan   e) Three Gorges Dam

5. Which of the following is not located in the area of the Pearl Estuary?

    a) Pudong   b) Hong Kong   c) Macau
    d) Xianggang   e) Shenzhen

6. Which of the following islands is now under the political control of Russia?

    a) Hainan   b) Hokkaido   c) Taiwan
    d) Shikoku   e) the Kuriles

True-False

1. China is expected to surpass the population total of 2 billion by the year 2010.

2. Despite industrialization, Japan still has a large agricultural sector, employing more than one-third of the labor force.

3. Both Hong Kong and Taiwan have land borders with China.

4. Since 1997 both Hong Kong and Macau have been returned to China's political control.

5. Mongolia forms the western anchor of the Jakota Triangle.

6. The communists' Long March terminated in the Huang (Yellow) Basin.

Fill-Ins

1. The leader of the legendary Long March through interior China was _____.

2. Japan's dominant economic region, which contains its largest city, is the _____ Plain.

3. The economic power that is destined to be under the political control of China after 1997 is _____.

4. The only Chinese city designated as an S.A.R. is _____.

5. The fastest-growing city in world history, which now plays a major economic role in the western Pacific Rim, is _____.

6. The legal concept whereby the property of one state lying within the boundaries of another forms an extension of the first, is known as _____.

Matching Question on East Asian Places

____1.	Tiananmen Square	A. Seoul
____2.	Nationalist Chinese capital	B. Nanjing
____3.	Dalai Lama's former domain	C. Edo
____4.	Bordering buffer state	D. Lantau
____5.	Chinese rustbelt city	E. Xinjiang
____6.	Special Economic Zone	F. Shanghai
____7.	South Korea's capital	G. Xizang
____8.	National belief system	H. Northeast China
____9.	Tokyo's former name	I. Shenyang
____10.	Capital of Taiwan	J. Shenzhen
____11.	One of Hong Kong's islands	K. Taipei
____12.	Home of the Manchus	L. Beijing
____13.	Japanese island	M. Mongolia
____14.	Junggar Basin	N. Shinto
____15.	Chang Delta	O. Hokkaido

Essay Questions

1. Discuss the key developments in the evolution of the Chinese state over the past 4000 years. Emphasize China's rich cultural heritage, explaining why Confucian (Kongzi) traditions still persist in many quarters despite the dominance of communism since 1949.

2. The North China Plain is one of the world's most densely settled areas. Discuss the environmental hazards of this subregion, its historical economic geography, and the impact that communism has had on the reorganization of its agricultural production.

3. Compare and contrast China's newest economic regions: the Special Economic Zones, Open Cities, and Open Coastal Areas. What is the purpose of each regional type, and show how each contributes to the Pacific Rim developments that are currently affecting much of China's seaboard.

4. Japan's record of modernization and industrialization over the past century is certainly one of the world's greatest national success stories. Trace the emergence of Japan as an industrial power, emphasizing resources, economic spatial organization, and international relations.

5. Define and discuss the economic tiger concept. Choose one of East Asia's economic tigers of the Pacific Rim, trace its recent development, and highlight its locational and economic-geographic advantages for such sustained national growth.

TERM PAPER POINTERS

The "Term Paper Pointers" section of the Introduction chapter in this **Study Guide** offered suggestions about approaching research and writing on geographic realms and their components, and you may wish to consult this material if you are undertaking a report on an East Asian region. You should also be aware of the textbook's Web Site (http://www.wiley.com/college/regions2000), which includes direct links to a number of other Web Sites which may be quite helpful to you in your research.

Several good treatments of East Asia's regional geography are listed in the **References and Further Readings** section of the book, and they are valuable references for regional analyses. Because of the closed nature of Chinese society and the inaccessibility of the People's Republic of China for Western geographers between 1949 and the early 1970s, this literature on China is not always as comprehensive as similar treatments of other world realms and regions. The most recent works on China are Buchanan et al., Cannon & Jenkins, Hsieh & Lu, Leeming, Pannell & Veek, C. Smith, Tregear, and Zhao (1994), with Pannell's edited volume also quite helpful. Spencer &

Thomas focus on the China of the mid-postwar period; both Ginsburg and Cressey, although classics, are now a generation out of date, basing much of their coverages on pre-communist China. For Japan see Burks, Harris, Kornhauser, MacDonald, Noh & Kimura, Reischauer, P. Smith, Trewartha, and Witherick & Carr; for Korea see Amsden, Cumings, and Hoare & Pares; for Taiwan see Ho and van Kemenade. An atlas well worth consulting is Hsieh & Hsieh. For treatment of post-1990 trends in China, the following are good sources: Barnett, Chapman & Baker, Chinoy, Dwyer, Harrell, Hook, Knapp, Kristof & WuDunn, LeHeron & Park, Lin, Mackerras, Pannell (1995), Qing, Shen, Spence, Starr, Terrill, van Kemenade, Veeck, Weiping, and Yeung (all titles)–and do not overlook Theroux's pithy but fascinating essay on South China's burgeoning development. Selya's wide-ranging geographic bibliography on China should also be consulted.

The systematic aspects of China's geography are generally more up-to-date in the literature. Environmental issues are examined in Hillel, Hoh, Hook, Qing, Smil (1984), and Zhao (both titles). Topics in cultural and historical geography are treated in Barnett, Harrell, Hsu & Serrie, Murphey, Salisbury, Spence, and Starr. Urban geography is surveyed in Kam, Lin, C.P. Lo, Lo & Yeung, Pannell (1995), Porter, Shen, Sit, Theroux, and Yeung (all titles). Economic geography, including population and regional development, is treated by Cook, Goodman, Jingzhi, Kornhauser, Lin, Linge & Forbes, McKay & van Grunsven, Potter et al., van Kemenade, Weiping, and Yeung (all titles). Topics in the systematic geography of Japan are treated by Burks, Cybriwsky, Harris, Karan & Stapleton, and P. Smith.

Much additional—and quite current—geographic literature on coastal East Asia can today be found under the rubric of Asia's western *Pacific Rim*. A good introduction to the geography of the western Pacific Rim can be found in Drakakis-Smith, Hodder, and Preston. More specialized works are those by Dirlik, Eccleston et al., LeHeron & Park, Lin, C.P. Lo, Lo & Yeung, McKay & van Grunsven, Murphy, Rumley et al., Theroux, van Kemenade, Vogel, and Yeung & Hu.

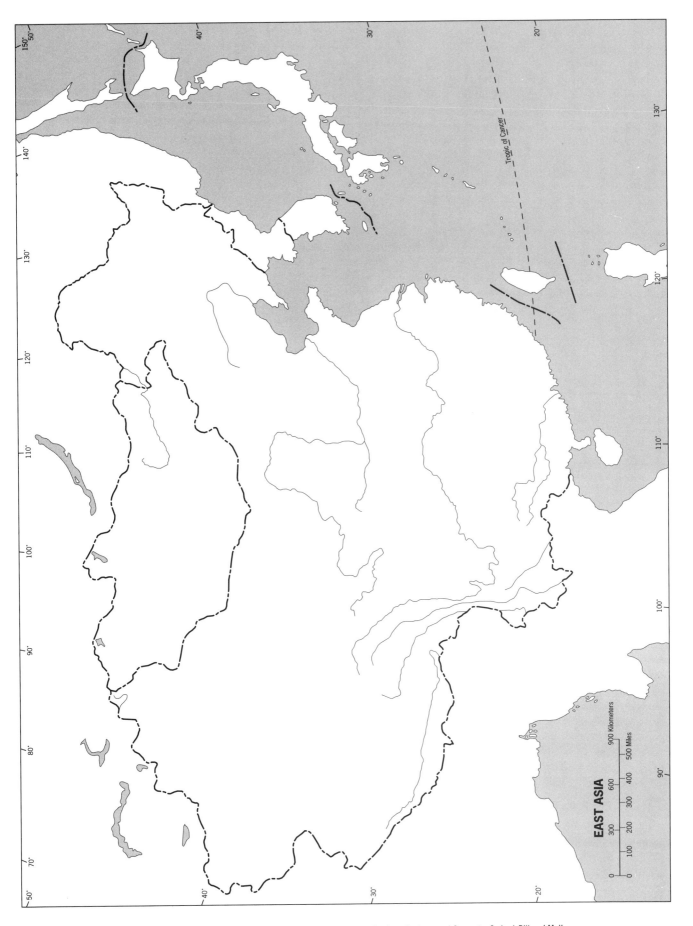

EAST ASIA

300   200   100   0
600   400   300   200   100   0
900 Kilometers
500 Miles

Tropic of Cancer

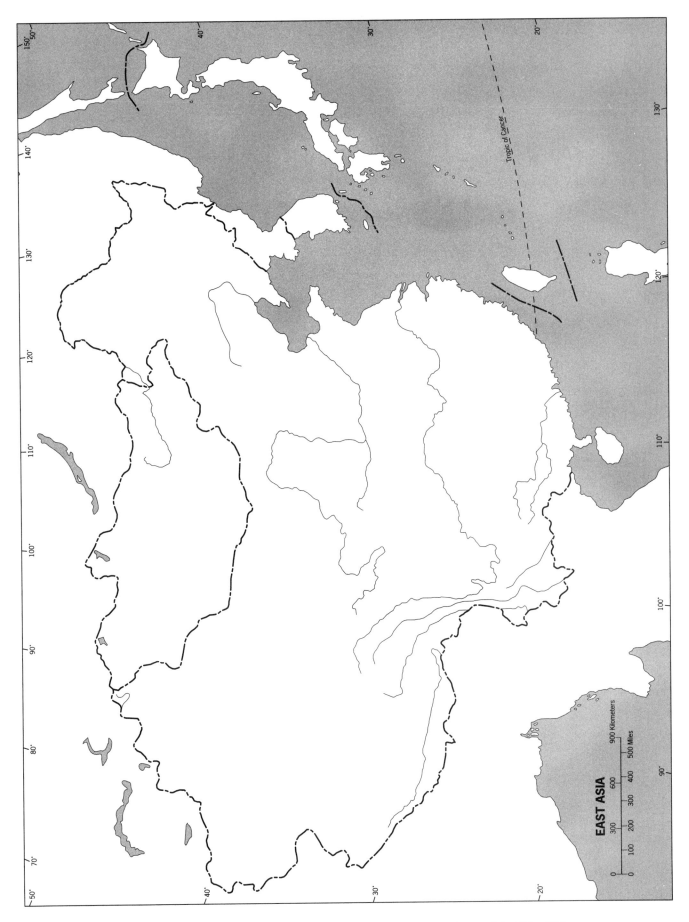

EAST ASIA

Tropic of Cancer

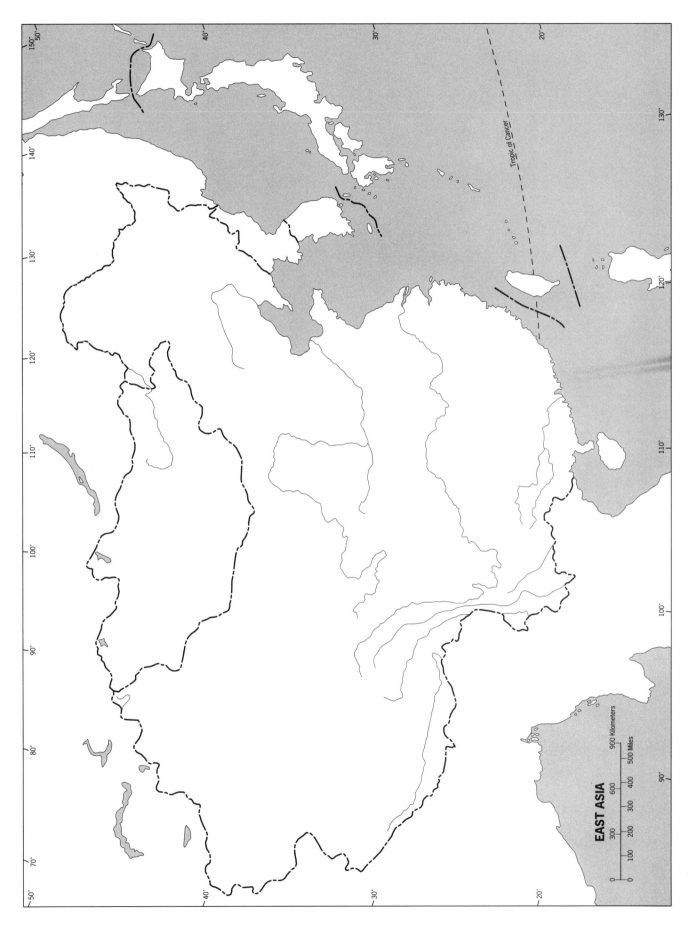

EAST ASIA

0 100 200 300 400 500 600 900 Kilometers

0 100 200 300 400 500 Miles

Tropic of Cancer

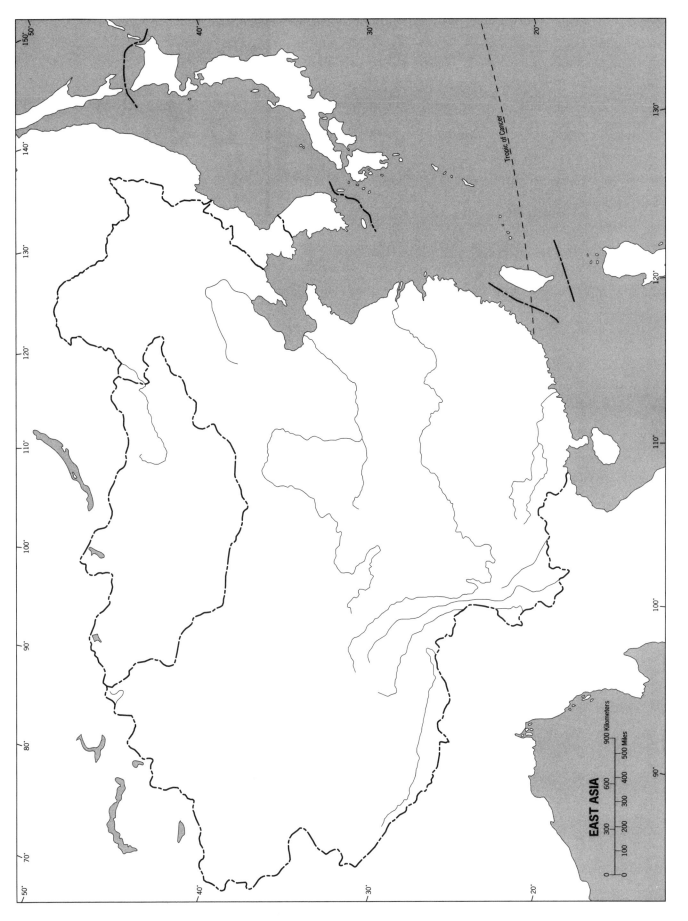

EAST ASIA

0	100	200	300	400		600		900 Kilometers
0			300		500 Miles			

Tropic of Cancer

CHAPTER 10
SOUTHEAST ASIA

OBJECTIVES OF THIS CHAPTER

Chapter 10 covers Southeast Asia, a realm of such cultural diversity, resulting from centuries of political contests by more powerful neighbors and European as well as Asian outsiders, that it is a classic example of a shatter belt. Following the introduction, population patterns are treated, and the subdivisioning of the realm into mainland and insular regional components is reviewed. This sets the stage for an overview of European colonial frameworks, highlighting the experiences of the British, the Dutch, and other European countries—including a review of the Chinese experience, which in some ways parallels colonialism. Territorial morphology is discussed at length, and provides the framework for the treatment of the realm's states.

Having learned the regional geography of Southeast Asia, you should be able to:

1. Understand the basic concepts of political geography, especially boundaries as they apply to the realm.

2. Describe the population distributions of the realm's mainland, peninsular, and island components.

3. Appreciate Southeast Asia's cultural-spatial fragmentation and the complex ethnic patterns that have evolved.

4. Grasp the diverse impacts that European colonialism shaped in this part of the world.

5. Understand the various categories of shape with respect to a state's territory, and the consequences of such spatial morphologies on the political development of Southeast Asia's states.

6. Locate the major physical, cultural, and economic-spatial features of the realm on an outline map.

GLOSSARY

Buffer zone *(494)*

A set of countries separating ideological or political adversaries. In southern Asia, Afghanistan, Nepal, and Bhutan were parts of a buffer zone between British and Russian-Chinese imperial spheres. Thailand was a buffer state between British and French colonial domains in mainland Southeast Asia.

Shatter belt *(494)*

Region caught between stronger, colliding external cultural-political forces, under persistent stress and often fragmented by aggressive rivals. Eastern Europe and Southeast Asia are classic examples.

Political geography *(504)*

The study of the spatial dimensions and expressions of political behavior.

Definition *(505)*

In boundary-making, a treaty-like document describing in words the location of an international border.

Delimitation *(505)*

The drawing of a defined boundary on an official map with the approval of the states being divided by that line.

Demarcation *(505)*

The actual placing of a boundary on the landscape in the form of a fence, cleared strip, or some other physical obstacle.

Physiographic boundary *(505)*

A boundary that coincides with features of the natural environment, such as rivers or mountains.

Anthropogeographic boundary *(505)*

A boundary that marks the break or transition between cultural groups.

Geometric boundary *(505)*

Straight-line or curved boundary.

Antecedent boundary *(506)*

A boundary defined and delimited before the main elements and settlement patterns of the present cultural landscape began to develop.

Subsequent boundary *(506)*

One that developed contemporaneously with the evolution of the major spatial elements of the cultural landscape.

Superimposed boundary *(506-507)*

A boundary placed by powerful outsiders on a developed human landscape, usually ignoring pre-existing cultural-spatial patterns.

Relict boundary *(507)*

One that has ceased to function but whose imprints are still visible on the cultural landscape.

State territorial morphology *(507)*

The size and shape of a state, and what that means in national political life.

Compact state *(507)*

One possessing roughly circular territory in which the distance from the geometric center to any point on the boundary exhibits little variation; Cambodia is a good example in Southeast Asia.

Protruded state *(507)*

One possessing territory that is at least in part a narrow, elongated land extension protruding from a more compact core area; in Southeast Asia, the southernmost portions of both Thailand and Myanmar (Burma) are examples.

Elongated state *(507)*

An attenuated state consisting entirely of territory that is at least six times longer than its average width; in Southeast Asia, Vietnam is a classic example.

Fragmented state *(507)*

One whose territory consists of several separate, non-contiguous parts, often isolated from one another by international waters or even the land areas of other states; Malaysia, Indonesia, and the Philippines are examples in Southeast Asia.

Perforated state *(507)*

One that completely encloses another state and is therefore perforated by it; most are small enclaves (such as the Vatican which perforates Italy)—the largest example is Lesotho within South Africa.

Domino theory *(510-box)*

The belief that political destabilization in one state can result in the collapse of order in a neighboring state, triggering a chain of events that, in turn, can affect a series of contiguous states.

Entrepôt *(520)*

A place, usually a port city, where goods are imported, stored, and transshipped; a *break-of-bulk* point.

Archipelago *(521)*

A set of closely grouped islands, usually elongated into a chain.

Transmigration *(526)*

Policy of the Indonesian governmental to encourage relocation of population from centrally-located and overcrowded Jawa to the more distant and less populated islands of the country. This aims to reduce population pressure and extend political strength to outlying areas.

SELF-TESTING QUESTIONS

Cover the right side of the page with a sheet of paper. Uncover each line after you have attempted to answer the question in the left column. If necessary, refer to the textbook page(s) listed at the right.

Question	Answer	Page
Realm Characteristics		
Why is Southeast Asia a shatter belt?	Its cultural-spatial fragmentation results from the collision of stronger outside powers within the realm, much like Eastern Europe.	494-496
Name the major population concentrations of this realm.	The basins of the Irrawaddy, Chao Phraya, Mekong, and Red Rivers; the fertile volcanic Indonesian island of Jawa; the plantation-studded western coast of the lower Malay Peninsula.	496

What percentage of Southeast Asia's population resides in Indonesia and the Philippines?	More than 55 percent of this realm's population is concentrated in these two countries.	496
Describe the realm's population distribution. Why is it different from the rest of Asia's?	Its leading population clusters are comparatively smaller and lie separated by wide expanses of sparse settlement. Physical obstacles are widespread, and agricultural opportunities are much more limited.	496-497

Southeast Asia's Political Geography

What were the major colonies of the British, French, Dutch, Spanish, and Portuguese?	The British held Burma (now Myanmar), Malaya, and northern Borneo; the French ruled Indochina; the Dutch dominated the East Indies (now Indonesia); as Fig.10-4 also shows, the Spanish held the Philippines and the Portuguese a part of eastern Indonesia.	499-501
Describe the origin of the modern state of Malaysia.	Upon independence in 1963, a federation emerged that included Singapore (which severed its ties in 1965); today this state consists of West Malaysia (peninsular Malaysia) and East Malaysia (Sarawak and Sabah on Borneo).	499
What does the term Indochina refer to?	The mainland quadrant of Southeast Asia that the French held in colonial times–the "Indo" referring to the cultural imprints from South and Southeast Asia (Hinduism, Buddhism, Indian art and architecture); "China" signifying the role of the Chinese in this realm.	501
Describe the distribution of the Chinese in Southeast Asia.	As the map on text p. 502 shows, there is a wide range of penetrations, peaking in western Malaysia, Jawa, the Mekong Delta, western Thailand, Singapore, and the northern portions of Vietnam, Myanmar, and the Philippines.	501-503

Political Geography

What idea underpins organic theory of state development?	Nations, since they are aggregates of human beings, would live and then die in the long run.	504
What are *centripetal* and *centrifugal* politico-spatial forces?	Centripetal forces unify the state; centrifugal forces are divisive.	504; ch.1, 56-box

What is meant by the boundary-related terms of *definition, delimitation,* and *demarcation*?	Definition is the document describing the boundary in words; delimitation is its placement upon an official map; demarcation is the actual placement of the border on the landscape.	505
What are the identifying characteristics of *antecedent, subsequent, superimposed,* and *relict boundaries*?	Antecedent boundaries precede cultural landscape development; subsequent boundaries emerge together with the elements of the cultural landscape; superimposed boundaries are forced by outsiders with little regard for existing cultural patterns; relict boundaries no longer function, but are still visible on the landscape.	506-507
What is meant by the term *state territorial morphology*?	The size and especially the shape of a country, and what they mean in national political life.	507
What are the identifying characteristics of *compact, fragmented, protruded,* and *elongated* state territorial morphologies?	Compact states are roughly circular in shape; fragmented states are split into two or more parts, often separated by international waters; protruded states exhibit long, narrow extensions; elongated states are at least six times longer than their average widths.	507

Regions of the Realm

What countries do the mainland and insular regions of Southeast Asia contain?	Mainland region: Vietnam, Cambodia, Laos, Thailand, Myanmar (Burma); Insular region: Malaysia, Singapore, Indonesia, Brunei, and the Philippines.	508
List the chief centrifugal politico-geographical forces for the following countries: Vietnam, Cambodia, Myanmar, Malaysia, and Indonesia.	Vietnam was constituted by the separate culture areas of Tonkin, Annam, and Cochin China, and suffered a generation of brutal warfare (from the mid-1940s to the mid-1970s), but is now reunited under communism. Despite Khmer cultural unity, Cambodia has been brutally treated by competing communist factions since 1975; Myanmar is beset by the problems associated with a brutal military government; Malaysia and Indonesia both suffer the effects of politico-spatial fragmentation that include considerable internal cultural variations.	508-513; 517-519; 525-526
Why has Singapore become an economic tiger?	It has benefitted from its relative location on the Strait of Malacca at the tip of the Malay Peninsula; large amounts of crude oil from Southwest Asia are refined; it is a major area of import/export trade; high-tech industries should help to secure a bright future.	520-521

What is the major religion of Indonesia?	Islam—in fact, constituting the world's largest Muslim country; but the faith is taken more casually here, and the islands east of Jawa are dominated by Hindus, Protestants, and Roman Catholics.	525
Where was the Chinese invasion felt most strongly in the Philippines?	In the northern part of the archipelago, especially on the largest island of Luzon.	527
Which religion is dominant in the Philippines?	Roman Catholicism, accounting for more than 80 percent of the population.	528
What are the three areas of densest population in the Philippines?	(1) The northwestern and south-central part of Luzon; (2) the southeastern extension of Luzon; (3) the islands of the Visayan Sea, between Luzon and Mindanao. See also Fig. I-9; p. 21.	529

MAP EXERCISES

Map Comparison

1. The importance of the Chinese in this realm has been amply demonstrated in the text. Compare Fig. 10-5 (p. 502) to the map of Southeast Asia's core areas (Fig. 10-2, p. 494), and discuss the spatial correspondence between the distribution of ethnic Chinese and the realm's key political decision-making centers.

2. The cultural diversity of Southeast Asia is readily apparent in the map of the realm's ethnic mosaic (Fig. 10-3, p. 498). Compare and contrast this map to the distribution of religions within Southeast Asia (see Fig. 6-2, pp. 284-285), making observations about the leading regional patterns of ethnic groups and religions.

3. Review the characteristics of antecedent boundaries on text p. 506. Then go back through all of the previous chapters in the text in search of other antecedent boundaries on appropriate maps. Make a list: you should be able to find numerous examples of such international boundaries.

Map Construction   (Use outline maps at the end of this chapter)

1. In order to familiarize yourself with Southeast Asian physical geography, place the following on the first outline map:

   a. *Rivers*: Irrawaddy, Salween, Mekong, Chao Phraya, Red

   b. *Water bodies*: Indian Ocean, Pacific Ocean, Andaman Sea, Java Sea, Flores Sea, Timor Sea, Banda Sea, Molucca Sea, Celebes Sea, Sulu Sea, Philippine Sea, South China Sea, Gulf of Thailand, Tonkin Gulf, Strait of Malacca, Sunda Strait, Tonle Sap

c. *Land bodies*: Spratly Islands, Paracel Islands, Hainan Island, Tonkin Plain, Kra Isthmus, Malay Peninsula, Jawa, Sumatera, Borneo, Sulawesi, Bali, Timor, Flores, Sumbawa, Molucca (Muluka) Islands, Mindanao, Luzon, Panay, Samar, Leyte, Palawan, Cebu, Visayan Islands

d. *Mountains and plateaus*: Annamese Cordillera, Khorat Plateau, Shan Plateau, Arakan Mountains, Yama Range, Dawna Range, Barisan Mountains, Iran Mountains, Muller Mountains

2. On the second map, political-cultural information should be entered as follows:

a. Label each country and its leading components (as appropriate).

b. Locate and label each capital city with the symbol *.

c. Using appropriate color pencils, reproduce the ethnic map (Fig. 10-3, p. 498), but for the Chinese instead of the color zones on this map use the darker-pink-colored areas shown in Fig.10-5, p.502.

3. On the third outline map, economic-urban information should be entered as follows:

a. *Cities* (locate and label with the symbol ●): Yangon (Rangoon), Mandalay, Moulmein, Bangkok, Thon Buri, Chiang Mai, Songkhla, Singapore, Kuala Lumpur, Pinang, Kelang, Alor Setar, Johor Baharu, Kuching, Bandar Seri Begawan, Kota Kinabalu, Banjarmasin, Samarinda, Jakarta, Surabaya, Bandung, Padang, Medan, Palembang, Surakarta, Malang, Pontianak, Ujung Pandang, Manado, Ambon, Phnom Penh, Siem Reap, Viangchan, Louangphrabang, Saigon-Cholon (Ho Chi Minh City), Hanoi, Da Nang, Hué, Loc Ninh, Haiphong, Manila, Cebu, Davao

b. *Economic regions* (identify with circled letter):

A - Irrawaddy Delta
B - Mekong Delta
C - Tonkin Plain
D - Chao Phraya Delta
E - Sarawak
F - Sabah
G - Kalimantan
H - Brunei

PRACTICE EXAMINATION

Short-Answer Questions

Multiple-Choice

1. A boundary delimited before a cultural landscape develops along its route is known as:

   a) antecedent
   d) superimposed
   b) demarcated
   e) subsequent
   c) relict

2. Reunited Vietnam's capital (still overwhelmingly called Saigon) is officially named after the communist leader who founded modern North Vietnam, a revolutionary named:

   a) Kota Kinabalu
   d) Dien Bien Phu
   b) Pol Pot
   e) Mao Zedong
   c) Ho Chi Minh

3. The largest Muslim country in the world in terms of population numbers is:

   a) Egypt
   d) Indonesia
   b) Indochina
   e) India
   c) Saudi Arabia

4. Singapore gained independence in 1965 by seceding from:

   a) the Malaysian Federation
   c) the Dutch East Indies
   e) the Association of Southeast Asian States
   b) the Sunda Archipelago
   d) the British Empire

5. Which of the following is not located on the island of Borneo?

   a) Sarawak
   d) Brunei
   b) Jakarta
   e) Sabah
   c) Kalimantan

6. Which of the following countries is not touched by the Mekong River?

   a) Laos
   d) Myanmar
   b) Vietnam
   e) Thailand
   c) Cambodia

True-False

1. The city commanding access to the strategic Strait of Malacca is Bangkok.

2. The dominant religion of both Indonesia and Malaysia is Islam.

3. Malaysia is an example of a fragmented state.

4. Singapore's population is dominated by ethnic Chinese.

5. There is no example of a perforated state in Southeast Asia.

6. Jakarta is located on the island of Sumatera.

Fill-Ins

1. The ethnic group that constitutes over 75 percent of Singapore's population is the ___.

2. Both Thailand and Myanmar are examples of states whose territorial morphologies can be classified as _____.

3. The most important island of the Philippines, which contains the capital of Manila, is called _____.

4. The capital of Vietnam is _____.

5. The easternmost island containing Indonesian territory is _____.

6. The religion originally associated with Cambodia's Angkor Wat temples was _____.

Matching Questions on Southeast Asian Countries

_____ 1. Most populous Muslim country   A. Brunei
_____ 2. Chinese majority   B. Philippines
_____ 3. Northeastern neighbor of Thailand   C. Cambodia
_____ 4. Tagalog speakers   D. Vietnam
_____ 5. Islamic sultanate   E. Myanmar
_____ 6. Angkor Wat ruins   F. Singapore
_____ 7. Chao Phraya core area   G. Indonesia
_ __ 8. Irrawaddy Delta   H. Laos
_____ 9. Contains Mekong Delta   I. Thailand
_____ 10. Sarawak and Sabah   J. Malaysia

Essay Questions

1. The population distribution of Southeast Asia differs markedly from that of the other Asian realms. Compare Southeast Asia's population pattern to India's and China's, discussing the physical, cultural, and economic geographic forces that account for the differences.

2. Review the course of colonialism in Southeast Asia, comparing and contrasting the British, French, and Dutch experiences, and highlighting the politico-geographical patterns that the colonial era has bequeathed to the present.

3. Define the notion of state territorial morphology and discuss its application to Cambodia, Malaysia, and Thailand, respectively, as *compact*, *fragmented*, and *protruded* states.

TERM PAPER POINTERS

The "Term Paper Pointers" section of the Introduction chapter in this **Study Guide** offered suggestions about approaching research and writing on geographic realms and their components, and you may wish to consult this material if you are undertaking a report on a Southeast Asian region.

You should also be aware of the textbook's Web Site, which includes direct links to a number of other Web Sites that may be helpful in your research. It is: (http://www.wiley.com/college/regions2000),

As always, the best way to begin is by getting hold of a good regional geography, especially for a realm whose cultural and environmental diversity is as complex as Southeast Asia's. Several are listed in the **References and Further Readings** section near the back of the book. Dixon, Dwyer, Dutt, and Rigg (both titles) are the most recent surveys, and Dutt, Fisher, Fryer, Hill, and Spencer & Thomas are fine older sources. The atlas by Ulack & Pauer should not be overlooked. More localized coverage is provided by Chapman & Baker, Cleary & Eaton, Cox, Dixon & Drakakis-Smith, Karnow, Kim et al., Leinbach & Sien, Murray & Perera, Neher & Marlay, Parnwell & Bryant, Schmidt et al., and Wurfel & Burton.

A number of systematic geographical studies are well worth considering. The historical-cultural aspects of this realm are treated by Broek, Burling, Hsu, Karnow, Murphey, Reynolds, SarDesai, and Schmidt et al. Political geography is discussed in Anderson, Cook et al., Glassner, "Good Fences," Hartshorne, Neher & Marlay, Prescott, and Rumley et al. Economic geography and development are covered in Brookfield et al., Cleary & Eaton, Dixon, Dixon & Drakakis-Smith, Kim et al., Leinbach & Sien, Neher & Marlay, Reynolds, and Wing-Kai et al.; the more specific topic of Pacific Rim-related development is treated in Cook et al., Drakakis-Smith, Eccleston et al., McKay & van Grunsven, Preston, Rumley et al., and Wing-Kai et al. Urban geography is examined by Ginsburg et al., Leinbach & Ulack, McGee, and Wing-Kai et al.. Environmental geography is treated in Brookfield et al. and Parnwell & Bryant.

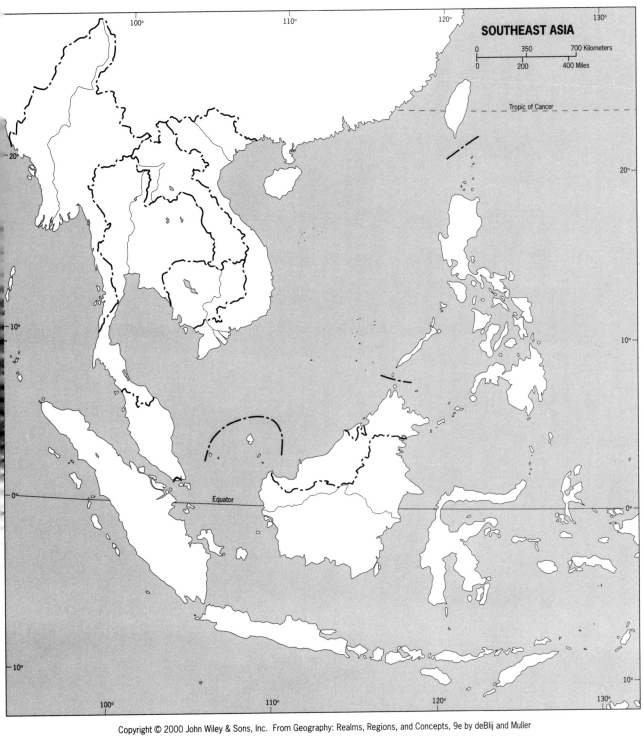

SOUTHEAST ASIA

0	350	700 Kilometers
0	200	400 Miles

Tropic of Cancer

Equator

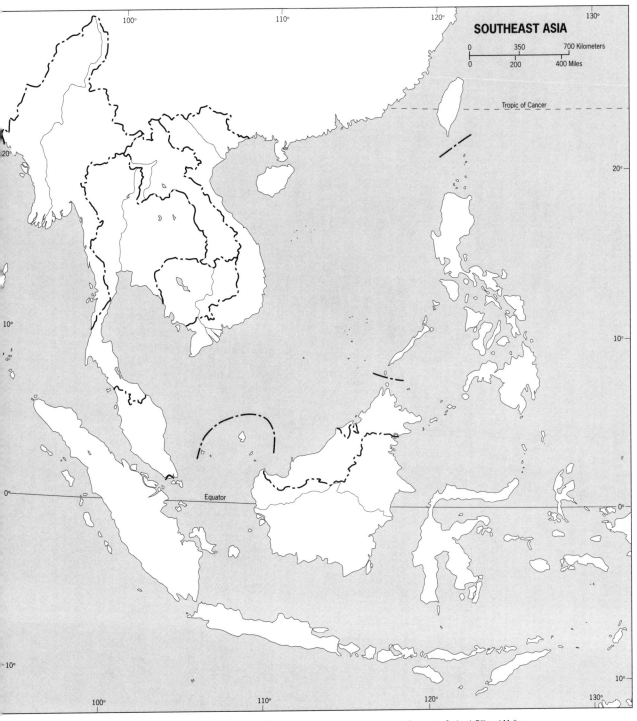

SOUTHEAST ASIA

0   350   700 Kilometers

0   200   400 Miles

Tropic of Cancer

100°   110°   120°   130°

20°

10°

Equator

0°

10°

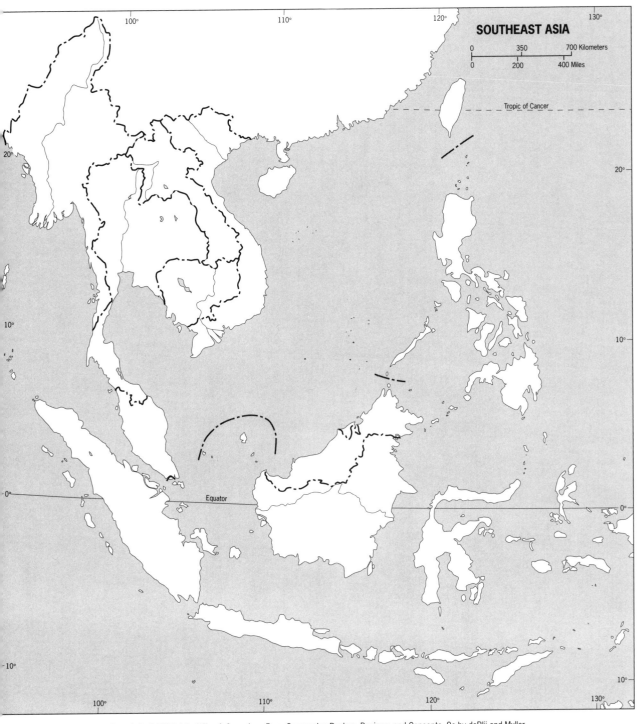

SOUTHEAST ASIA

0	350	700 Kilometers
0	200	400 Miles

Tropic of Cancer

Equator

100° 110° 120° 130°

20°

10°

0°

-10°

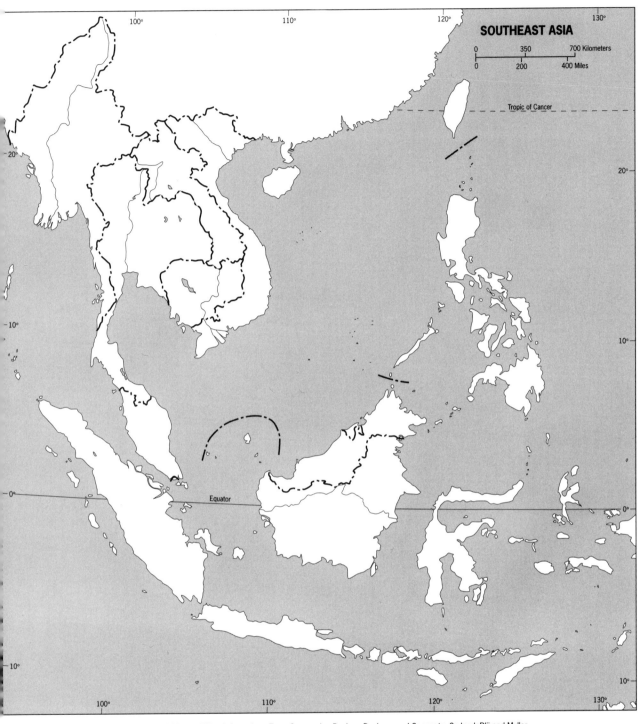

SOUTHEAST ASIA

| 0 | 350 | 700 Kilometers |
| 0 | 200 | 400 Miles |

Tropic of Cancer

Equator

CHAPTER 11
THE AUSTRAL REALM

OBJECTIVES OF THIS CHAPTER

We now move "downunder" to survey Australia and its neighbor, New Zealand. An introduction to the realm's natural environment includes a brief description of Australia's unique biogeography. Australia's federal system is then explored in context of the country's modern evolution. The economic geography of Australia is examined next: productive activities, exports, economic problems, and the functional status of the economy within the sphere of the western Pacific Rim are some of the issues discussed. Population patterns are then covered, as well as Australia's immigration policies. A discussion of New Zealand's geography ends the chapter, with a description of the country's physical features, economic activities, and ethnic tensions.

Having learned the regional geography of Australia and New Zealand, you should be able to:

1.   Understand Australia's political geography and why federalism has succeeded there.

2.   Understand the course of Australia's modern evolution, with special emphasis on the country's economic geography.

3.   Note the key population patterns and immigration issues of Australia.

4.   Understand the main points of New Zealand's human geography.

5.   Locate the leading physical, cultural, and economic-spatial features of Australia and New Zealand on a map.

GLOSSARY

Austral *(534)*

Geographic term meaning "south".

Subtropical convergence *(535-box)*

A narrow marine transition zone, girdling the globe at approximately latitude 40° S, that marks the equatorward limit of the frigid Southern Ocean and the poleward limits of the warmer Atlantic, Pacific, and Indian Oceans to the north.

Biogeography *(536)*

The study of fauna (animal life)and flora (plant life) in spatial perspective.

Ecosystem *(536)*

The interactions between a group of organisms and their natural environment.

Aboriginal population *(536-537)*

Native peoples, especially in Australia. Often used to designate the inhabitants of areas that were conquered and subsequently colonized by the imperial powers of Europe.

Outback *(539)*

The name given by Australians to the vast, peripheral, sparsely-settled interior of their country.

Federation *(540)*

Derived from the Latin word *foederis*, meaning "association." A political framework wherein a central government represents the various subnational entities (Australia's 6 states and 2 territories in this case) within a nation-state where they have common concerns—defense, foreign affairs, and the like—yet permits these entities to retain their own laws, policies, and customs in certain spheres.

Unitary state *(540)*

Deriving from the Latin word *unitas*, meaning "one." A unified and centralized state, usually with a long tradition of being ruled by a single authority, in which power is exercised equally throughout the country.

Import-substitution industries *(542)*

Industries set up by local entrepreneurs in and near market centers in a remote country (such as Australia) because import costs of foreign goods are extremely high.

Environmental degradation *(546)*

The accumulated human abuse of a region's natural landscapes that, among other things, can involve air and water pollution, threats to plant and animal ecosystems, misuse of natural resources, and generally upsetting the balance between people and their habitat.

SELF - TESTING QUESTIONS

Cover the right side of the page with a sheet of paper. Uncover each line after you have attempted to answer the question in the left column. If necessary, refer to textbook page(s) listed at the right.

Question	Answer	Page
Biogeography		
What does biogeography study?	It focuses on fauna (animal life) and flora (plant life) in spatial perspective.	536
Who was one of the founders of biogeography as a systematic field?	Alfred Russel Wallace, who did extensive research in Australia.	536
What two fields is biogeography divided into?	Phytogeography—focusing on plants; zoogeography—focusing on animals.	536
Australia's Dimensions		
Is Australia part of the Pacific Rim?	Only spatially. Functionally, Australia can be likened to an open-pit mine of raw materials and agricultural exports oriented to the booming economies to the north of it. Its manufactures are not high-tech finished products, unlike those of many western Pacific Rim countries.	537
Where is most of Australia's population concentrated?	In the east and southeast, mostly facing the Pacific Ocean. A second, smaller core focuses on the southwest. These regions coincide with the non-desert, humid temperate climate zones.	539

What is meant by the term *Outback*?	The vast, dry, sparsely populated interior of Australia.	539
When did Australia's indigenous peoples arrive on the island continent?	Aboriginal Australians arrived some 50,000-60,000 years ago, developing a patchwork of indigenous cultures.	539

Australia's Political Geography

Why is Australia a federal state?	Because it was formed as a union composed of several separate colonies who wanted to preserve their identities and customs.	539-540
When was the Australian Federation formed?	The Commonwealth of Australia was created in 1901–it celebrates its centennial in 2001.	539-540

Australia's Urban and Economic Geography

What percentage of to-day's Australians reside in cities or towns?	Fully 85 percent.	540
How is Australia functionally organized?	Like Japan, large cities are located along the coast, and are the foci for agriculture and manufacturing.	540
What are some of Australia's economic problems?	The price of farm products, a major Australian export, fluctuates on the world market; expensive petroleum imports are needed; government-protected industries are backed by strong labor unions; inflation has grown, the national debt rose, and unemployment grew.	542
Name Australia's three leading agricultural exports.	Wool, meat, and wheat.	542
Where are Australia's main dairying zones located?	Close to the large urban markets, as in other areas of the world.	542
What kind of spatial pattern is exhibited by Australia's varied mineral deposits?	Scattered across the country, but in high enough concentrations to make mining profitable. See Fig. 11-5.	542-543

What is the present condition of Australian manufacturing?	Concentrated in urban areas; quite diversified; domestically oriented.	543-544

Australia's Population

Are Australia's European ties as strong as they used to be?	No, a new age is dawning in Australia; its Asian ties are steadily strengthening.	544, 544-box
What percentage of the Australian population now is of British/Irish origins?	Only about 33 percent—down from about 75 percent in the 1970s.	545
Why is the Australian government under pressure to limit immigration once again?	It is a time of economic difficulty, and immigration (mainly from Asian countries) was at an all-time high in the 1990s.	545-546

New Zealand

Who are the pre-European people of New Zealand, Polynesians who are still present in large numbers?	The Maori.	547
How do Australia's and New Zealand's landscapes differ?	While Australia's land is of generally low relief and elevation, New Zealand's is generally quite high and has very rugged relief.	547
What is the chief farming region of the South Island?	The Canterbury Plain.	548

MAP EXERCISES

Map Comparison

1. Compare the map of Australian minerals and agricultural areas (Fig.11-5, p.543) with the country's population distribution (Fig. I-9, p. 21). What spatial associations are apparent, and for which activities is remoteness a serious problem?

2. Compare the maps of Australia and New Zealand's population/settlement patterns (Fig. 11-4, p. 538 and 11-7, p. 547) to both the map of world climates (Fig. I-8, pp. 14-15) and Australia's physiography (Fig. 11-2, p. 535). What spatial relationships can be discerned? Are Australia and New Zealand's patterns of settlement similar or different?

Map Construction (Use outline maps at the end of the chapter)

1. To familiarize yourself with Australia and New Zealand's physical geography, place the following on the first of the outline maps:

   a. *Water bodies*: Murray River, Darling River, Bass Strait, Great Australian Bight, Gulf of Carpentaria, Coral Sea, Tasman Sea, Cook Strait

   b. *Land areas*: Great Dividing Range, Murray Basin, Nullarbor Plain, Great Artesian Basin, Great Sandy Desert, Great Victoria Desert, Arnhem Plateau, Gibson Desert, Tasmania, Southern Alps, Canterbury Plain, North Island, South Island

2. On the second outline map, enter the following politico-geographical information:

   a. draw in all international borders on the map and label each country.

   b. label all of the States and Territories of Australia, and label each of their capitals with the symbol □; label the national capital with the symbol *.

3. On the third outline map, enter the following urban-economic information:

   a. *Cities* (locate and label with the symbol ●) : Sydney, Melbourne, Adelaide, Brisbane, Canberra, Hobart, Newcastle, Rockhampton, Alice Springs, Darwin, Perth, Kalgoorlie, Wellington, Auckland, Christchurch, Dunedin

   b. *Economic centers* (identify with a circled letter):

      A-   Broken Hill
      B-   Kambalda
      C-   Mount Isa
      D-   Coronation Hill
      E-   Canterbury Plain

PRACTICE EXAMINATION

Short-Answer Questions

Multiple Choice

1. Australia's governmental structure can best be classified as a:

    a) unitary state   b) federal state   c) dictatorship
    d) monarchy   e) republic

2. The group that reached Australia 50,000-60,000 years ago is known as the:

    a) Aborigines   b) Indians   c) British
    d) Chinese   e) Marsupials

3. Which city located is located on Australia's western, Indian Ocean coast:

    a) Sydney   b) Auckland   c) Melbourne
    d) Perth   e) Canberra

4. Australia's federal form of government was established in:

    a) 1492   b) 1788   c) 1901
    d) 1927   e) 1976

True-False

1. There was a gold rush in the 1850s in Australia.

2. New Zealand's population, like Australia's, is largely concentrated on its coastal fringes.

3. The Australian Outback is a rugged, highland zone.

4. Australia's percentage of urban population is higher than that of the United States.

Fill-Ins

1. Australia's largest city is _____.

2. New Zealand's larger island is its _____ Island.

3. New Zealand's capital is _____.

4. The capital of Victoria State and Australia's second largest city is _____.

Matching Question on Australia/New Zealand

_____ 1. Australia's only large, non-coastal city   A. Maori
_____ 2. New Zealand's largest lowland   B. Canterbury Plain
_____ 3. Largest minority in New Zealand   C. Auckland
_____ 4. Sugarcane-growing area of Australia   D. Queensland
_____ 5. New Zealand's largest city   E. Canberra
_____ 6. Capital of South Australia   F. Adelaide

Essay Questions

1. Australia is only slowly reorienting itself to become a major player in western Pacific Rim developments. Discuss why this is the case, including such factors as exports, general economic conditions in the country, and its locational situation. Explore these same factors for the leading countries of the western Pacific Rim, such as Japan, Singapore, and Taiwan, and make comparisons.

2. There is a growing risk of ethnic polarization in New Zealand. Describe the Maori situation—its historical development, current status, and prospects for the future.

TERM PAPER POINTERS

The "Term Paper Pointers" section of the Introduction chapter in this **Study Guide** offered suggestions about approaching research and writing on geographic realms and their components, and you may wish to consult this material if you are undertaking a report on Australia/New Zealand. You should also be aware of the textbook's Web Site (http://www.wiley.com/college/regions2000), which includes direct links to a number of other Web Sites that may be quite helpful in your research.

The best general sources on Australia in the current geographic literature are Barrett & Ford, Heathcote (1994), Jeans (both volumes), and McKnight (1995). Two older works worth consulting are Spate and McKnight (1970). The atlas by Camm & McQuilton is a useful compendium of historical maps. More popular overviews are offered by Bambrick, and Terrill. A number of good systematic works are available because geography is a leading academic discipline in Australia. Economic geography is the focus in Courtenay, Day & Rowland, "Environment and Development," Heathcote (1988), Heathcote & Mabbutt, Hodder, Rich, Walmsley & Sorenson, and Williams. Population geography topics are examined in Day & Rowland, Head, Heathcote & Mabbutt, Inglis, Smith, and Walmsley & Sorenson. Urban geography is treated in Hofmeister and Statham. Political geography is examined by Rumley. Historical geography is covered in Camm & McQuilton, Head, Hughes, Lines, Meinig, and Powell. Environmental topics are treated in "Environment and Development," Gentilli, Heathcote (1988), Heathcote & Mabbutt, Lines, and Serventy; the more specific field of biogeography is examined in Beadle, Darlington, and Parkes. New Zealand is treated in Britton, Cumberland & Whitelaw, Franklin, Holland & Johnston, R.J. Johnston, and Morton & Johnston.

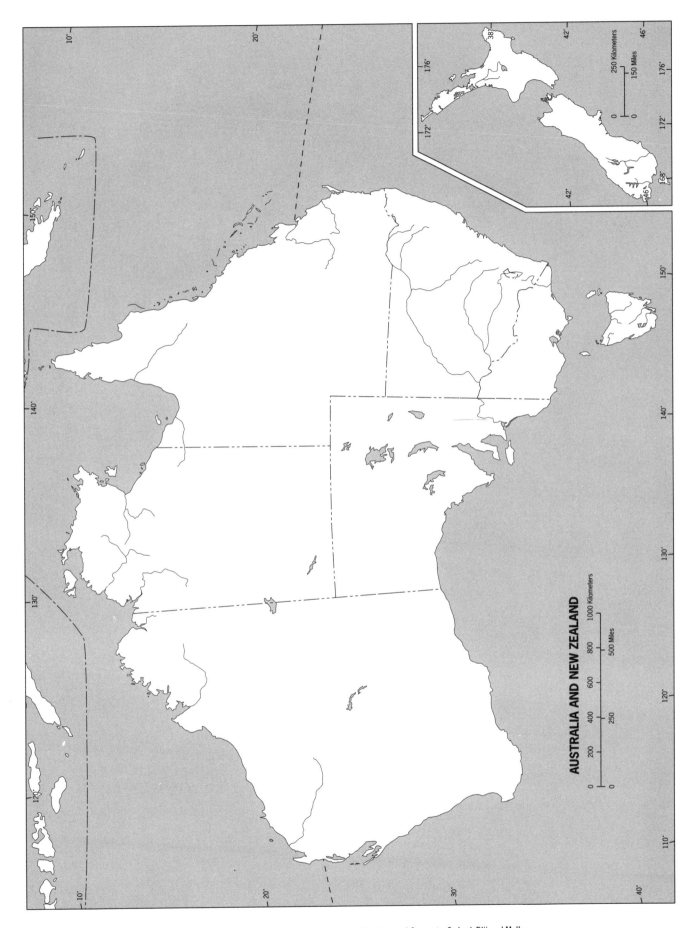

AUSTRALIA AND NEW ZEALAND

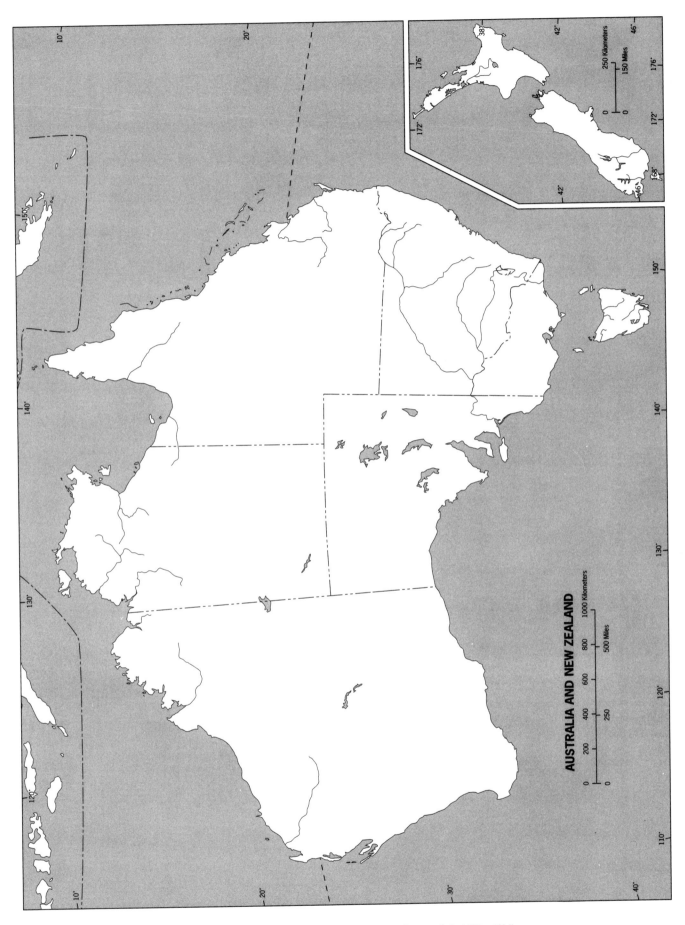

AUSTRALIA AND NEW ZEALAND

0   200   400   600   800   1000 Kilometers
0   250   500 Miles

0   250 Kilometers
0   150 Miles

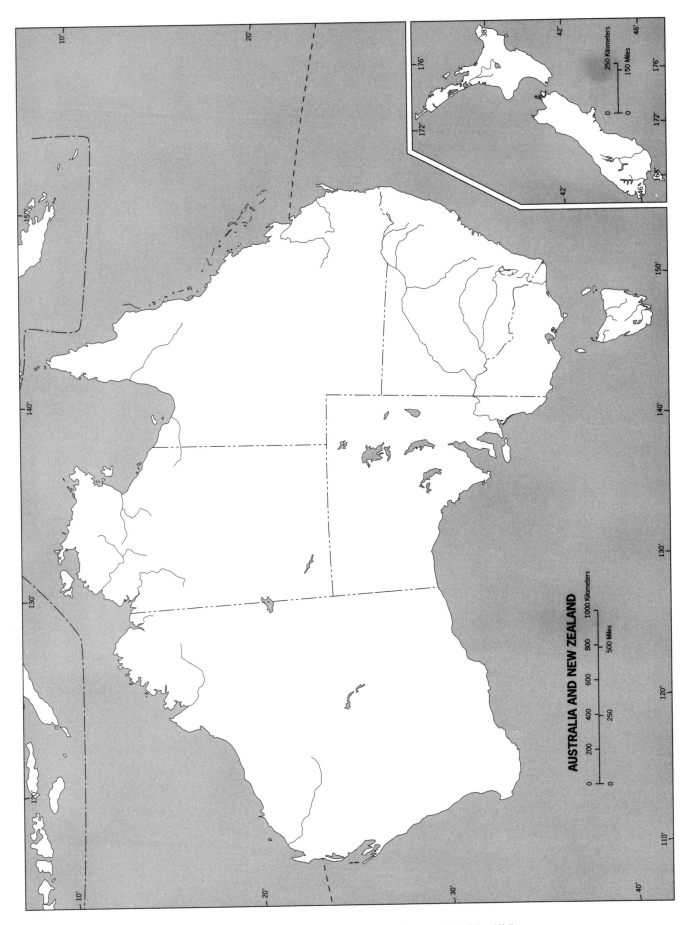

AUSTRALIA AND NEW ZEALAND

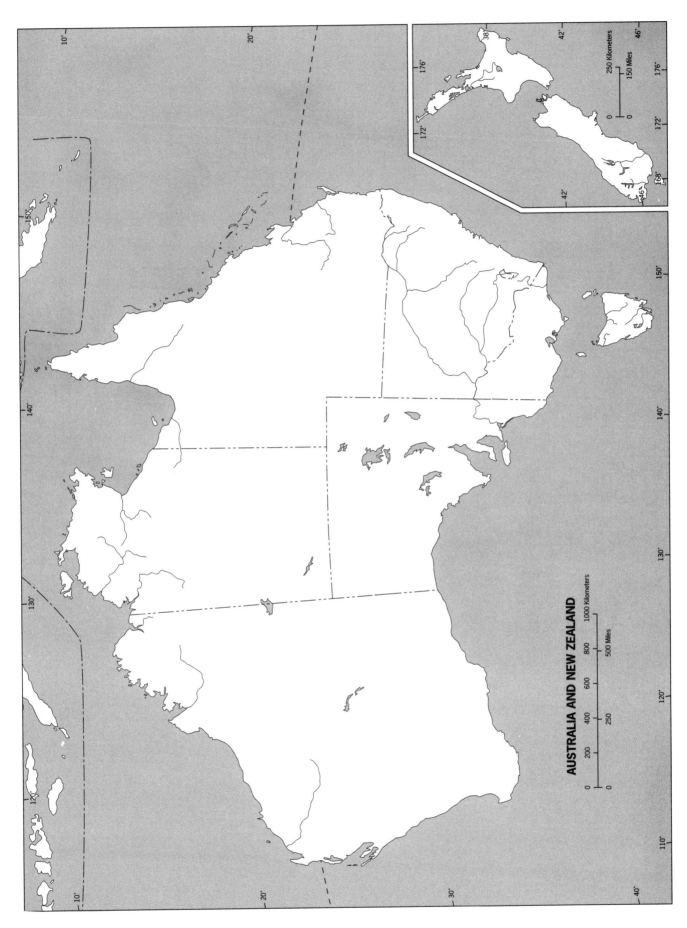

AUSTRALIA AND NEW ZEALAND

CHAPTER 12
THE PACIFIC REALM

OBJECTIVES OF THIS CHAPTER

This final chapter covers the largest area by far, greater than any of the previously-treated realms—the vast Pacific Ocean, whose thousands of islands contain inhabitants who dwell in a fragmented, highly complex geographic realm. Appropriately, the topic of marine geography is introduced, and this is followed by an elaboration of the political geography of maritime claims. We then turn to a survey of the realm's three regions—Melanesia, Micronesia, and Polynesia.

Having learned the regional geography of the Pacific Realm, you should be able to:

1. Comprehend basic concepts of marine geography and maritime claims.

2. Appreciate its fragmented cultural complexities as well as its regional commonalities.

3. Describe the leading geographic characteristics of Melanesia, Micronesia, and Polynesia, and locate their major places on an outline map.

4. Understand this unique realm's opportunities and challenges given its diverse components.

GLOSSARY

Marine geography *(553)*

The geographic study of oceans and seas. Its practitioners investigate both the physical (e.g., coral-reef biogeography, ocean-atmosphere interactions, coastal geomorphology) and human aspects (e.g., maritime boundary-making, fisheries, beachside development) of oceanic environments.

Territorial sea *(553)*

The water adjoining a country's coastline which is under the legal control of that state.

High seas *(553)*

Ocean areas free from the legal claim of any state, because they are beyond the defined Exclusive Economic Zones.

Continental shelf *(553)*

The shallow, sloping seafloor that leads away from a coastline down to about 100 fathoms (600 feet).

Exclusive Economic Zone (EEZ) *(553)*

An oceanic zone extending up to 200 nautical miles from a shoreline, within which the coastal state can control fishing, mineral exploration, and additional activities by all other countries.

Maritime boundary *(553)*

An international boundary that lies in the ocean. Like all boundaries, it is a vertical plane, extending from the seafloor to the upper limit of the air space in the atmosphere above the water.

Median line boundary *(553)*

An international maritime boundary drawn where the width of the sea is less than 400 nautical miles. Because the states on either side of that sea claim Exclusive Economic Zones of 200 nautical miles, it is necessary to reduce those claims to a (median) distance equidistant from each shoreline. Delimitation on the map almost always appears as a set of straight-line segments that reflect the configurations of the coastlines involved.

Melanesia (558-560)

The most populous Pacific region covering New Guinea and the island groups stretching to its southeast (see Fig. 12-3, pp. 556-557).

Micronesia (560-561)

The chains of small islands lying north of Melanesia and east of the Philippines (see Fig. 12-3, pp. 556-557).

High islands (560)

Volcanic islands of the Pacific Realm that are high enough in elevation to wrest substantial moisture from the tropical ocean air. They tend to be well watered, their volcanic soils enable productive agriculture, and they support larger populations than low islands– which possess none of these advantages and must rely on fishing and the coconut palm for survival.

Low islands (560)

Low-lying coral islands of the Pacific Realm that–unlike high islands–cannot wrest sufficient moisture from the tropical ocean air to avoid chronic drought. Thus productive agriculture is impossible and their modest populations must rely on fishing and the coconut palm for survival.

Polynesia (561-562)

The remainder of the realm, lying within the great triangle circumscribed by New Zealand, Hawai'i, and Easter Island (see Fig. 12-3, pp. 556-557).

Claimant state (563)

A state with a territorial claim to an area; in this case refers to Antarctica (see Fig. 12-4, p. 562).

Cover the right side of the page with a sheet of paper. Uncover each line after you have attempted to answer the question in the left column. If necessary, refer to textbook page(s) listed at the right.

Question	Answer	Page
Marine Geography		
What politico-spatial problems do marine geographers study?	How far a state's jurisdiction extends off its coastline; by what method seaward boundaries are defined, delimited, and sometimes demarcated; and what responsibilities states have in the remaining "high seas."	553
What is a state's territorial sea?	The strip of water adjoining a country's coastline, which is legally controlled by that state.	553
What was the outcome of the UNCLOS conferences?	Among key provisions, a 12-mile territorial sea was defined for all countries; a 200-mile Exclusive Economic Zone, in which states have exclusive rights, was also created.	553
Why was it necessary to draw median-line boundaries between many countries' territorial seas?	Many nations are located next to choke points, and thus are separated from each other by less than 24 miles of water. Both 12-mile territorial seas and 200-mile Exclusive Economic Zones would overlap, creating endless maritime boundary disputes.	553
Dimensions of the Pacific Realm		
When did Papua New Guinea become a sovereign state?	In 1975, after nearly a century of British and Australian administration.	558
What are the two largest cultural groups of Papua New Guinea?	The largest group is the Papuans, who live in the south; the Melanesians are second in terms of population size, and inhabit the north and east of the country.	558
Describe the economic potential of Papua New Guinea.	Oil was discovered in the 1980s, and became the largest export by value in the 1990s. Gold, silver, and many other materials are now exported, as well as agricultural products, largely to Pacific Rim countries.	558
What does the term *Micronesia* mean?	It is derived from the tiny average size of its islands (*micro* means small).	560

Differentiate between *high-island* and *low-island* cultures.	High-island cultures are associated with volcanic islands whose fertile soils permit farming; low-island cultures are based on dry, low-lying coral islands, and forced to rely on fishing.	560
What does the term *Polynesia* mean?	It is derived from the large number of islands it contains (*poly* means many) within its vast areal extent, as shown in Fig. 12-3, pp. 556-557.	561
Describe the current political geography of Polynesia.	Highly complex. The United States' 50th state comprises the Hawaiian archipelago; Tuvalu, Kiribati, and Tonga are former British dependencies; certain islands remain under French control, including Tahiti; other islands continue to be administered under the flags of Chile, New Zealand, the U.S., and the United Kingdom.	562
Do Antarctica and the Southern Ocean constitute a geographic realm?	In physiographic terms, yes, but in functional terms, no. No functional regions, cities, transport networks, etc. have developed.	562
Why are states interested in making territorial claims in remote Antarctica?	Land and sea contain raw materials, such as fuels, minerals, and proteins—which may become crucial reserves in the future.	563

MAP EXERCISE *(There are no map exercises for this chapter.)*

Even though there is no outline map (which would be impractical at the one-page scale), the major places to learn can be listed:

Islands: New Guinea, Bougainville, Solomon Islands, New Caledonia, Hawaiian Islands (Oahu, Hawai'i, Maui, Molokai, Lanai, Kauai), Guam, Midway Island, Wake Island, Nauru, Fiji, Tonga, Samoa, Tahiti, Easter Island, Pitcairn Island.

Water bodies: Coral Sea, Arafura Sea, Philippine Sea.

Political Units: Papua New Guinea, Solomon Islands, Vanuatu, New Caledonia, Wallis and Futuna, Western Samoa, American Samoa, Tonga, Fiji, French Polynesia, Tokelau Islands, Nauru, Kiribati, Tuvalu, Cook Islands, Easter Island, Niue, Rarotonga, Truk, and—comprising the former U.S. Trust Territory in Micronesia—the Northern Mariana Islands, Republic of Palau, Federated States of Micronesia, and Republic of the Marshall Islands.

Cities: Honolulu, Hilo, Port Moresby, Suva, Pago Pago, Nouméa, Papeete.

PRACTICE EXAMINATION

Short-Answer Questions

Multiple-Choice

1. The major economic activity associated with low-island cultures is:

    a) hunting   b) farming   c) fishing

    d) tourism   e) nomadic herding

2. Which Melanesian island, still ruled by France today, is one of the world's largest sources of nickel?

    a) New Caledonia   b) New Guinea   c) Tahiti

    d) Fiji   e) Bougainville

True-False

1. Papua New Guinea is located in Polynesia.

2. High-island cultures are associated with fertile soils and farming economies.

Fill-Ins

1. The most heavily populated of the three Pacific regions is _____ .

2. The largest in areal extent of the three Pacific regions is _____ .

Matching Question on the Pacific Realm

_____	1. Micronesia's largest island	A. Port Moresby
_____	2. Large-scale copper mining	B. Kiribati
_____	3. Capital of Papua New Guinea	C. Bougainville
_____	4. Large majority of Hawaiians located here	D. Oahu
_____	5. Formerly called the Gilbert Islands	E. Guam

Essay Questions

1. Compare and contrast the major geographical dimensions of Melanesia, Micronesia, and Polynesia, and evaluate the potential for future development and modernization in each region.

2. Discuss the conflict on the island of Bougainville. Describe the centrifugal forces at work, noting the history of Papua New Guinea's political development, and comment on future prospects for this state.

TERM PAPER POINTERS

The "Term Paper Pointers" section of the Introduction chapter in this **Study Guide** offered suggestions about approaching research and writing on geographic realms and their components, and you may wish to consult this material if you are undertaking a report on a Pacific region. You should also be aware of the textbook's Web Site, which includes direct links to other Web Sites which may be quite helpful in your research. It is: (http://www.wiley.com/college/regions2000).

In the **References and Further Readings** section of the textbook, the best general surveys are provided by Brookfield (1973), McKnight, and Peake. The excellent and quite current map by Bier is an important cartographic reference on all parts of this realm; Karolle performs a similar function for Micronesia. Other useful overviews of the Pacific Basin are offered by Bunge & Cook, Carter, Couper (*Development*), Freeman, "Mobility and Identity," Oliver, Sager, and Theroux. Individual regions are covered in Brookfield & Hart, Connell, Damas, Goodman et al., Grossmann, Howard, Howlett, Karolle, Kirch, Kluge, and Leibowitz.. Hawai'i is covered by Morgan (both titles) and Woodcock; Antarctica is treated in Chaturvedi, Crossley, de Blij, Dodds, and Sugden. A number of systematic topics are also worth considering. Historical geography is the focus of Friis, McEvedy, Spate (all titles), and Ward. Population geography is examined in Grossmann, "Mobility and Identity," Peake, Quanchi, Robillard, Vayda, and Ward. Economic geography and development topics are the focus of Connell, Couper (*Development*), Damas, Goodman et al., Grossmann, Kirch, Kissling, Lockhart et al., and Nunn. Environmental dimensions are treated in Mitchell and Zurick. Marine geography (including maritime political geography) is treated in Blake (both titles), Cole, Couper (*Atlas*), Glassner, Johnston & Saunders, Nunn, and Prescott.

ANSWERS TO
PRACTICE SHORT-ANSWER EXAMINATIONS

Introduction
Multiple-Choice:	1(c), 2(b), 3(a), 4(a), 5(e), 6(c)
True-False:	1(true), 2(true), 3(true), 4(false), 5(true), 6(false)
Fill-Ins:	1(India), 2(functional), 3(B), 4(cultural landscape), 5(New Zealand), 6(relative)
Matching:	1(B), 2(I), 3(H), 4(D), 5(F), 6(A), 7(G), 8(J), 9(L), 10(C), 11(E), 12(K)

Chapter 1
Multiple-Choice:	1(d), 2(a), 3(d), 4(d), 5(d), 6(a)
True-False:	1(true), 2(true), 3(true), 4(true), 5(true), 6(false)
Fill-Ins:	1(von Thünen), 2(Spain), 3(centripetal), 4(devolution), 5(Denmark), 6)Cyprus
Matching:	1(J), 2(F), 3(N), 4(P), 5(K), 6(I), 7(B), 8(M), 9(O), 10(D), 11(L), 12(E), 13(H), 14(A), 15(G), 16(C)

Chapter 2
Multiple-Choice:	1(d), 2(a), 3(e), 4(c), 5(d), 6(c)
True-False:	1(false), 2(false), 3(false), 4(true), 5(true), 6(true)
Fill-Ins:	1(*glasnost*), 2(Volga), 3(Alaska), 4(centrifugal), 5(St. Petersburg), 6(Kuzbas)
Matching:	1(D), 2(F), 3(A), 4(G), 5(H), 6(J), 7(B), 8(C), 9 (E), 10(I)

Chapter 3
Multiple-Choice:	1(d), 2(a), 3(b), 4(c), 5(a), 6(c)
True-False:	1(true), 2(true), 3(true), 4(true), 5(true), 6(false)
Fill-Ins:	1(Boston), 2(Cascade), 3(St. Lawrence), 4(Northern Frontier), 5(Western Frontier), 6(Nunavut)
Matching:	1(G), 2(I), 3(K), 4(M), 5(O), 6(B), 7(D), 8(C), 9(H), 10(A), 11(J), 12(L), 13(E), 14(F), 15(N)

Chapter 4
Multiple-Choice:	1(d), 2(a), 3(e), 4(a), 5(c), 6(e)
True-False:	1(true), 2(false), 3(false), 4(true), 5(true), 6(false)
Fill-Ins:	1(Spain), 2(*caliente*), 3(Mexico), 4(Cuba), 5(Honduras), 6(Panama)
Matching:	1(O), 2(K), 3(L), 4(N), 5(C), 6(F), 7(B), 8(M), 9(E), 10(J), 11(D), 12(H), 13(G), 14(A), 15(I)

Chapter 5
Multiple-Choice:	1(b), 2(a), 3(c), 4(c), 5(d), 6(b)
True-False:	1(false), 2(false), 3(false), 4(false), 5(false), 6(false)
Fill-Ins:	1(copper), 2(CBD), 3(*templada*), 4(São Paulo), 5(Patagonia), 6(Bolivia)
Matching:	1(E), 2(D), 3(K), 4(G), 5(M), 6(L), 7(J), 8(H), 9(B), 10(C), 11(A), 12(I), 13(F)

Chapter 6

Multiple-Choice:	1(c), 2(a), 3(d), 4(b), 5(c), 6(b)
True-False:	1(false), 2(true), 3(true), 4(false), 5(true), 6(false)
Fill-Ins:	1(Cyprus), 2(Relocation), 3(Shi'ah or Shi'ite), 4(Jordan), 5(Cairo), 6(Caucasus)
Matching:	1(K), 2(L), 3(E), 4(D), 5(O), 6(M), 7(F), 8(N), 9(C), 10(I), 11(A), 12(H), 13(G), 14(J), 15(B)

Chapter 7

Multiple-Choice:	1(e), 2(a), 3(e), 4(e), 5(a), 6(d)
True-False:	1(false), 2(true), 3(true), 4(true), 5(false), 6(false)
Fill-Ins:	1(Nigeria), 2(separate development), 3(malaria), 4(Cape Town), 5(The Congo), 6(Madagascar)
Matching:	1(H), 2(G), 3(I), 4(D), 5(J), 6(K), 7(A), 8(E), 9(C), IO(F), 11(L), 12(B)

Chapter 8

Multiple-Choice:	1(d), 2(c), 3(a), 4(c), 5(e), 6(b)
True-False:	1(false) 2(true), 3(false), 4(false), 5(true), 6(false)
Fill-Ins:	1(caste), 2(Coromandel), 3(physiologic), 4(Sinhalese), 5(Arabian), 6(Nepal)
Matching:	1(F), 2(J), 3(L), 4(H), 5(I), 6(G), 7(A), 8(K), 9(E), 10(C), 11(D), 12(B)

Chapter 9

Multiple-Choice:	1(b), 2(a), 3(b), 4(b), 5(a), 6(e)
True-False:	1(false), 2(false), 3(false), 4(true), 5(false), 6(true)
Fill-Ins:	1(Kanto), 2(Mao Zedong), 3(Hong Kong [Xianggang]), 4(Hong Kong [Xianggang]), 5(Shenzhen, China), 6(extraterritoriality)
Matching:	1(L), 2(B), 3(G), 4(M), 5(I), 6(J), 7(A), 8(N), 9(C), 10(K), 11(D), 12(H), 13(O), 14(E), 15(F)

Chapter 10

Multiple-Choice:	1(a), 2(c), 3(d), 4(a), 5(b), 6(d)
True-False:	1(false), 2(true), 3(true), 4(true), 5(true), 6(false)
Fill-Ins:	1(Chinese), 2(protruded), 3(Luzon), 4(Hanoi), 5(New Guinea), 6(Hinduism)
Matching:	1(G), 2(F), 3(H), 4(B), 5(A), 6(C), 7(I), 8(E), 9(D), 10(J)

Chapter 11

Multiple-Choice:	1(b), 2(a), 3(d), 4(c)
True-False:	1(true), 2(true), 3(false), 4(true)
Fill-Ins:	1(Sydney), 2(South), 3(Wellington), 4(Melbourne)
Matching:	1(E), 2(B), 3(A), 4(D), 5(C), 6(F)

Chapter 12

Multiple-Choice:	1(c), 2(a)
True-False:	1(false), 2(true)
Fill-Ins:	1(Melanesia), 2(Polynesia)
Matching:	1(E), 2(C), 3(A), 4(D), 5(B)